PHYSICAL DISABILITIES - THERAPEUTIC IMPLICATIONS

Edited by **Uner Tan**

Physical Disabilities - Therapeutic Implications
http://dx.doi.org/10.5772/65159
Edited by Uner Tan

Contributors

Masayoshi Wada, Yu Munakata, Lin-Ju Kang, Ai-Wen Hwang, Chia-Ling Chen, Carmen Fernández-Panadero, Begoña Polonio-López, Marta Rodríguez-Hernández, Olga López-Martín, Stanislava Klobucka, Elena Ziakova, Miroslav Malay, Jana Havlova, Nina Sladekova, Jaroslav Kresanek, Gurusidheshwar Mahadevappa Wali, Gautam Wali, Zohreh Salimi, Martin Ferguson-Pell, Pablo Aqueveque, Esteban Pino, Francisco Saavedra, Enrique Germany, Britam Gomez, Paulina Ortega, Yuki Iida, Kunihiro Sakuma, Pooja Kulkarni, Alok Sharma, Hemangi Sane, Nandini Gokulchandran, Prerna Badhe, Suhasini Pai, Ritu Varghese, Amruta Paranjape

Published by InTech
Janeza Trdine 9, 51000 Rijeka, Croatia

Publishing Process Manager Martina Blecic
Technical Editor SPi Global
Cover InTech Design team

First published June, 2017
Printed in Croatia
Legal deposit, Croatia: National and University Library in Zagreb

Additional hard copies can be obtained from orders@intechopen.com

Physical Disabilities - Therapeutic Implications, Edited by Uner Tan
p. cm.
Print ISBN 978-953-51-3247-9
Online ISBN 978-953-51-3248-6

For EU product safety concerns:
IN TECH d.o.o., Prolaz Marije Krucifikse Kozulić 3, 51000 Rijeka, Croatia,
info@intechopen.com or visit our website at intechopen.com.

Meet the editor

Prof. Dr. Uner Tan (born May 1, 1937) is a Turkish neuroscientist, best known for his discovery and study of human quadrupedalism with cognitive decline (Uner Tan syndrome). He taught at various universities in Germany and Turkey. Tan is the honorary member of the Turkish Academy of Sciences and won numerous awards including the Science Award from the Turkish Scientific and Technical Council, Einstein and Nobel Medals for Science and Peace (Albert Einstein Foundation, USA), and Gold Record for Brain Research (American Biographical Institute). Tan's scientific studies include the cerebrospinal motor systems of humans and animals, walking gaits, cerebral lateralization, intelligence, handedness, finger-length patterns, and Uner Tan syndrome (quadrupedalism and cognitive decline).

Contents

Preface

This book presents studies covering a wide range of physical disabilities, i. e., pathological conditions limiting the healthy physical activities. In other words, a physical disability may be defined as a physical impairment (partial or total) that causes difficulty or error in motor actions. Various aspects of this condition will be dealt with in the present book.

Chapter 1, by Zohreh Salimi and Martin W. Ferguson-Pell, "Validity in Rehabilitation Research: Description and Classification," considering all of the available classification data, provides an all-inclusive classification of different types of research validity. In addition, a review of the literature was also conducted.

Chapter 2, by Gurusidheshwar Wali and Gautam Wali, titled "Rare and Disabled Movement Disorders: An Indian Experience," presents Uner Tan syndrome cases with habitual quadrupedal locomotion and impaired cognition recently discovered in India. It also considers the transport defects in subcellular organelles in these cases, including the potential rectification with known drugs raising hopes for their cure.

In Chapter 3, "Five-Wheeled Wheelchair with an Add-On Mechanism and Its Semiautomatic Step-Climbing and Step-Descending Function," the authors, Masayoshi Wada and Yu Munakata, propose a novel add-on electric drive system for propelling a manual wheelchair on the floor together with an advanced function to climb and descend a step with no human support.

Chapter 4, "Participation and Environmental Factors of Children with Physical Disabilities in Taiwan," by Lin-Ju Kang, Ai-Wen Hwang, and Chia-Ling Chen, reports that "participation is a critical health and education outcome of children and can be optimized by environmental supports; children with physical disabilities often experience participation restriction and environmental barriers."

Chapter 5, by Marta Rodríguez-Hernández, Carmen Fernández-Panadero, Olga López-Martín, and Begoña Polonio-López, under the title "Hand Rehabilitation after Chronic Brain Damage-Effectiveness, Usability, and Acceptance of Technological Devices: A Pilot Study," deals with an overview of existing tools for hand rehabilitation after brain injury and a pilot study to test HandTutor in patients with chronic brain damage, concluding that this chapter is a step further to evaluating the acceptance of technological devices in patients with chronic brain damage.

Chapter 6, written by Nina Sladekova, Elena Ziakova, Jaroslav Kresanek, Stanislava Klobucka, Jana Havlova, and Miroslav Malay, is, as the title suggested, focused on the impact of non–robot-assisted therapy for paretic upper extremity caused by cerebral palsy, using Armeo Therapy. The authors compare it to classical kinesiotherapy.

Chapter 7, "After-Stroke Movement Impairments: A Review of Current Technologies for Rehabilitation," by Pablo Aqueveque, Paulina Ortega, Esteban Pino, Francisco Saavedra, Enrique Germany, and Britam Gomez, provides a review of the rehabilitation technologies for people who have suffered a stroke.

Chapter 8, written by Alok Sharma, Hemangi Sane, Nandini Gokulchandran, Prerna Badhe, Pooja Kulkarni, Suhasini Pai, Ritu Varghese, and Amruta Paranjape, deals with stem cell therapy in pediatric neurological disabilities. In the chapter, the use of novel monitoring tools such as MRI MSK and PET-CT scan to track the changes occurring at the cellular level after stem cell therapy is described.

Finally, Chapter 9, by Yuki Iida and Kunihiro Sakuma, presents a review of the literature regarding critical illness-induced muscle wasting and describes potential treatment of excessive muscle catabolism.

Acknowledgment

I am grateful to the Turkish Academy of Sciences (Ankara, Turkey) as well as InTech's publishing team for the support during the review process.

Prof. Dr. Uner Tan
Honorary Member of the Turkish Academy of Sciences, Ankara, Turkey
Cukurova University, Medical School, Department of Physiology, Adana, Turkey

Chapter 1

Validity in Rehabilitation Research: Description and Classification

Zohreh Salimi and Martin W. Ferguson-Pell

Additional information is available at the end of the chapter

http://dx.doi.org/10.5772/67389

Abstract

There is a considerable body of literature on research validity across different disciplines. With regard to rehabilitation research, however, this body is narrow and warrants further consideration. Classification of research validity has been considered and developed over the past six decades; however, a literature search returned no comprehensive discussion that has gathered all available classifications under an overarching umbrella. The aim of this chapter is to provide an all-inclusive classification of different types of research validity, focusing on rehabilitation research. A basic review of the body of literature available was conducted. Different classifications of validity in the literature were recognized, considered, unified, and are presented in this chapter. Moreover, the main threats to each type of validity and some strategies to minimize them are discussed. A classification of all types of validity in rehabilitation research is presented in this chapter. Furthermore, the matter of priority between these research validities is discussed. It is concluded that while all types of research validities are important to be considered, maximizing all of them in one research project is sometimes controversial. Thus, researchers should make a situation-based trade-off between different aspects of validity in order to optimize the overall validity of their research.

Keywords: research validity, construct validity, classification, threats to validity, rehabilitation research

1. Introduction

Getting concerned with the validity of a research is ensuring its empirical integrity [1]. The level of validity of a study is an indicator of cost-effectiveness and accountability of it [1]. A research study is considered valid when (1) it is able to correctly find the relationship between the variables, (2) it measures what it claims to be measuring, (3) the findings are generalizable, and also (4) it has adequate statistical power to reject a false null hypothesis. On the other hand, the power of a study is its ability to find the truth: correctly rejecting a false-null hypothesis or supporting a true-null hypothesis. Therefore, a valid study is also a powerful study, because its findings are the true results of that study.

All researchers need to ensure the validity of the tests and instruments they use before conducting research studies, but this is of particular importance for rehabilitation researchers: in a review study, 100 data-based studies were reviewed and by a *post hoc* power calculation, it was shown that there is a high possibility for the occurrence of type II error in rehabilitation research studies [1]. Type II error refers to supporting a false-null hypothesis, which leads to a false-negative conclusion. In the rehabilitation studies reviewed in this study, the medians of power for detecting small-, medium-, and large-effect sizes have been only equal to 0.08, 0.26, and 0.56, respectively. These low-power values clearly show the importance of accounting for power and validity of the study in rehabilitation research, before starting to conduct one. This is because the purpose of all studies is truly discovering the relationship among variables [2], and if a study does not have validity, the results obtained from it will not be trustworthy.

Validity of the research should be considered for all research studies. Research studies concerned with validity have been done for a large variety of different fields of knowledge, including Education [3–12], Psychology [13–16], Marketing [17], Management [18], Employment [19], Nursing [20], Criminal issues [21], Animal studies [22], Sports [23], Nutrition and Food Science [24], Health [25–27], and Rehabilitation Science [1, 2]. Regarding the number of research studies about validity that are available in the literature for different disciplines, Education, Psychology, and Marketing are the fields that are most prominent. However, the body of literature in this context for health studies is narrow, and even narrower for rehabilitation science. This highlights the necessity of addressing this issue for health sciences in general, and rehabilitation sciences in particular. In this chapter, the validity of different types of rehabilitation studies will be discussed. In doing so, the main focus will be on one type of validity that is a more complicated aspect of validation: construct validity. We will discourse some threats to each type of validity along with providing some strategies to prevent them, and hence how to power up the study, as we pass through each validation type.

2. What is validity?

Validity principles are applicable to all studies, whether they are based on questionnaires, observational studies, or other types of assessments [7]. Research validity helps us to know how true the claims and propositions made in a study are. This judgment can be based on the characteristics of a study, such as the research design, adequacy of sample size, the recruitment procedure, instruments and tests used, and the appropriateness of statistical methods used [28].

3. Types of research validity

Research validity can be categorized into two types: internal validity and external validity [29]. Internal validity refers to the amount of credit that can be attributed to the relationship between variables that is true, and external validity refers to how generalizable are the findings. In another approach [28], research validity has been divided into four types: internal validity, statistical conclusion validity, construct validity, and external validity. These four types of validity address four basic questions that practicing researchers face and so it is more practical. Those basic questions are as follows:

(1) Is there any relationship between two variables? (Referring to internal validity)

(2) If so, is it a causal relationship or it might just have happened by chance and can occur without any treatment? (Referring to statistical conclusion validity)

(3) What are the concepts that are involved in this causal relationship? (Referring to construct validity)

(4) How generalizable is this relationship to other settings, tools, persons, and time? (Referring to external validity)

The second classification is actually drawn by dividing each of the first classification components in two [28]: the statistical conclusion validity is differentiated from internal validity to distinguish between the relationship that is affected by covariates and the true relationship between the two variables that are of interest. In other words, internal validity takes care of the validity of the relationship obtained between the two main variables of interest, while the statistical validity makes sure that this relationship is not contaminated (and if so, is correctly taken care of) by other variables that may influence the relationship, but are not of the interest in the study.

Also, construct validity is differentiated from external validity to make a distinction between generalization of the constructs of cause and effect to higher-order constructs and generalization of the findings to the other settings and the population. In other words, the second classification explicitly states what was implicitly covered in the first classification.

The second classification of validity is now widely accepted and is being used by researchers in different fields [1, 2]. We will briefly introduce each of these types of validity in the following sections of this chapter. Furthermore, some threats to each type of validity as well as some strategies to power up that validity will be discussed.

3.1. Statistical conclusion validity

Statistical conclusion validity can be defined as the approximate precision of interpretations concluded about the covariations based on the statistical methods used and the fitness of the research design [2]. In other words, any statistical issue that could influence the results of the study would be a threat to statistical conclusion validity, including small sample size, lack of statistical power for the study [1], and any random error that could happen by chance, despite appropriate use of statistics, for example, type I and type II errors [30]. These threats to validity happen when the research conditions and statistical process of the study are not rigorous enough [22]. This type of validity is the *most important* type of validity, but in rehabilitation research, it has received little attention [1].

Some of the main threats to statistical conclusion validity are [2, 28] as follows:

- **Low statistical power:** This will reduce the power of the study to reject the null hypothesis. To prevent this threat, researchers should define their experiment characteristics, for example, sample size and eligibility criteria, in order to provide an acceptable statistical power for the research.
- **Violated statistical assumptions**: Each statistical procedure is based on some assumptions, for example, normal distribution of population, homogeneity of variance, and linearity. If these assumptions are violated, those statistical procedures will not be credible anymore. To prevent this threat, the researcher should ensure that the underlying assumptions are met, for example, normality of sample data.
- **Performing multiple statistical tests on a single data set**: This increases the likelihood of type I error and is another threat to statistical validity. This is because the Alpha level of the study as a whole is the sum of the Alpha level of all comparisons made. To prevent this threat, researchers can reduce the number of comparisons by a careful preplanning. Another technique is to determine the Alpha level of each *t*-test in a way that the cumulative Alpha would be desired.
- **Lack of inter- and intra-rater reliability of the study**: Lack of consistency in conditions of implementing experiments is a threat to validity.
- **Lack of reliability of measures:** This would result in both type I and type II errors (e.g., they might show no correlation between two variables, when there is a good correlation, and vice versa). To prevent this threat, researchers must only use reliable measures or tools. In fact, reliability is a necessary condition for validity [14].

3.2. Internal validity

Internal validity deals with whether the treatment used in the research study has an actual effect on the outcome variable [30]. Thus, internal validity is the extent to which we can be confident that there is a certain type of relationship (e.g., causal) between the dependent and the independent variables of the study [2]. If a study has lack of internal validity, then there are some factors in that study, other than the independent variables, that affect the outcome to some extent, but they have not been accounted for [2, 30]. These unaccounted factors are threats to internal validity. There are many threats to internal validity which can be applicable to different types of research. Some of the most common threats are [2, 28, 30] as follows:

- **Maturation:** When experiments get lengthy, the results may be influenced by the participants getting older, wiser, healthier, or stronger. Maturation is considered a threat to validity when it has not been considered in the research design and is not accounted for.
- **History:** When subjects' reaction in the experiment has been influenced by some events that have happened prior to the experiment, for example, when the participant is studied to observe his/her attitude to people with disabilities after some treatments, but in fact his/her past experiences of encountering a person with disability would affect the results.
- **Lack of inter- and intra-rater reliability** will also affect the results and so the causal relationship concluded from those results.
- **Selection**: The researcher should assure that what makes a difference between control and treatment groups is the only factor that is under study. Any other variables that might influence the outcome should not systematically differ between groups. The best protection against this threat is randomly assigning participants to treatment and control groups when possible.
- **Attrition and mortality:** Dropouts that usually happen most for the group that is receiving the harder treatment would influence the outcome, since those who comply with the treatment are generally those who are healthier or are more enthusiastic about the research study they have participated in. Also, mortality can potentially happen for every experiment that needs the attendance of participants for more than one time, but especially it may happen for studies where the participants have serious diseases, such as cancer or heart disease.
- **Sharing information**: If by any means participants have a chance to share their information regarding the experiment, they might influence each other's thoughts and outcomes. This is even more important when the experiment is based on questionnaires and other qualitative methods.

If participants know about other participants in a different group, they may get dissatisfied with the group that they are in and with the treatment type they are receiving and thus they may become disappointed and less motivated to appropriately continue with the treatment (e.g., performing some exercises), which may affect their compliance. Any protection against chances of getting cues about other participants will help to prevent sharing of data. Any strategies that can be taken against hypothesis guessing by participants and also performing a good concealment for the randomization will also contribute to this aim.

3.3. External validity

External validity refers to the population that the study findings can validly be generalized to. In other words, the greater the population the findings can be applied to, both in number and in diversity, the higher the external validity of that study. This, however, depends significantly on the sample used in the study and the population that this sample is actually representing. A simple example of for this is that if you are aiming for generalizing your findings to all wheelchair users but you only recruit male wheelchair users, you have a problem in external validity of your research. Also, the instruments used in the study and other conditions of the experiments have roles in determining the population the study findings can be generalized to and so in defining the external validity of the study.

This type of validity addresses the generalizability of the findings, which is of particular importance for rehabilitation practitioners, because they need to make inferences from the sample under the study to the treatment provided to a greater population [2]. For making those inferences, we either should perform the same experiment on in different occasions in time, settings, participants, and so on, or one should perform a systematic review with meta-analysis on the body of literature on a particular issue [30]. However, randomly selecting participants in a study (i.e., random sampling) will provide the best protection against threats to external validity of that study and therefore makes the findings generalizable to the population [2]. Some of the main threats to external validity are [2] as follows:

- **Sample characteristics** (e.g., age, gender, race, education, urban versus rural) restrict the population that the findings can be generalized to. Samples should be a good representative of the desired population. Random sampling helps considerably to this aim, but it does not guarantee that this threat would be eliminated.
- **Intervention characteristics** also restrict the findings to the settings with similar features, for example, instruments. Some strategies to prevent this threat include making the use of different examiners (when intra-rater reliability is realized) and using multiple measures that are taken from multiple setting.
- **Context characteristics:** There are some conditions that may influence the way subjects react or respond in the experiment, which inhibit generalization of the findings to the situations with different conditions. For instance, some participants try to provide "correct" responses, which are responses that they believe examiners like to see, but are not representative of their real state. Moreover, sometimes subjects receive multiple treatments at the same time, which may restrict the findings to those people that are on similar treatment regimes.
- **Sensitization:** In research designs where participants receive the same assessment (e.g., questionnaire) pre- and post-test, their knowledge on the way they will be evaluated might affect the outcomes. Since this situation might not be the same as when there is no sensitization on the construct under study, this is a threat to external validity.

3.4. Construct validity

For performing powerful studies and also proper clinical accomplishments, we need to make use of robust measures, considering the fact that "science rests on the adequacy of its measurement" [14]. Using proper and robust measures pertains to construct validity. Construct validity deals with whether an instrument or measurement tool or a test is measuring what it claims to be measuring [11]. In other words, construct validity concerns whether the measurement obtained is really representing the underlying construct [14].

The matter of validity is analogous to a study that has a clear hypothesis; researchers should gather as much evidence as they can to prove the hypothesis about validity of the inference

[11, 13, 14]. Researchers should continue gathering convincing evidence until they feel that they have a large enough set of evidence to prove the construct validity. There is no best way to validate a study, although there are several methods in use [11]. Up to four subclasses of construct validity have been defined: face validity, content validity, criterion-related validity, and construct validity. Some researchers [11, 14, 30, 31] believe that all these subclasses should be grouped and gathered under one overarching umbrella which is the construct validity. This is called the unified view of construct validity. Each of the subclasses of construct validity will be discussed below.

3.4.1. Face validity

This is the first judgment about the validity of an instrument by just looking at the appearance of it. It is only guesswork and provides little evidence for validity of the instrument. Besides, some snags can happen when talking about face validity [14]: the fallibility of verdicts that are based on appearance, different interpretations of appearance between developers and users, and some occasions that the judgment based on the appearance of an instrument is contrary to its contents. Therefore, using only face validity is never sufficient.

3.4.2. Content validity

This type of validity concerns the items or elements of a measurement and the extent to which these elements reflect the area they are supposed to be measuring [31, 32]. The adequacy and fitness of each element of the measurement tool in measuring the targeted construct is discussed under content validity. In other words, the targeted construct guides selecting the content of an assessment tool, and on the other hand the content and elements of the assessment tool selected define the construct that is actually being investigated [32]. Content validity is particularly important in assessing the validity of questionnaires.

Using an instrument that is invalid due to content, results in erroneous conclusions because some aspects of the construct are not represented properly, whether underrepresented or overrepresented. Not accounting for content validity in the study could also result in inaccuracy in finding a significant treatment effect [31].

Content validity is dynamic in nature, because the domain and definition of constructs change by time, and accordingly the elements of an instrument should be changed to be representative of that construct [31]. Content validity of an assessment instrument is dependent on the function of the instrument, population under study, and the situation in which the instrument is used. Therefore, close attention should be taken in order to maintain an acceptable validity for the assessment instrument. Often, a panel of experts is contacted to judge content validity of an assessment or instrument [11, 31].

One example of threats to content validity are is occasions where the definition of one term in researchers' language is different from the commonly accepted definition of it. In this situation, the readers' interpretations of the contents, results, and reports of the study might be different from the author's and researcher's intent. In these cases, a proper clarification of the constructs is advised [2].

3.4.3. Criterion-related validity

In criterion validity, correlation of the instrument or assessment with a "criterion" is examined [11]. Criterion has to be a "superior" measure that is more accurate than the measure being evaluated; otherwise, the failure in validation might be due to a flaw of the criterion, itself [32]. There are two types of criterion validity [11, 32]: concurrent and predictive, which are introduced briefly, here as follows:

- *Concurrent validity*: This pertains to situations that both the assessment tool that is being tested for its validation and the criterion are measured at the same time. For instance, when blood pressure is measured simultaneously using cuff measurements and intra-arterial pressure measurement tools [32].
- *Predictive validity:* When the assessment tool is tested for its validity by checking how well it can predict a criterion that will happen later. For instance, how well the scores of a test obtained by a sample of people predict their job status in the future [32]. Diagnosis, physiological data, and tests performed in laboratories are examples of instances that predictive validation should be used [32].

3.4.4. Construct validity

In construct validity, we experimentally investigate whether a construct is actually measuring what it claims to be measuring [11]. The concept of construct validity emerged when researchers realized there are many occasions that there is not any "superior" criterion to correlate the instrument under validation study with it [32]. Two types of validity can be distinguished within construct validity [32]:

- *Convergent validity*: Convergent validity is used when in validating a method of measurement, the correlation between that measure and a different method of measurement is assessed, while they are used to measure the same construct [33].
- *Discriminant (divergent) validity*: When we experimentally show that our assessment tool being tested for its validity produces results that are different from data produced by another assessment tool that is measuring another construct and thus should produce different results [33].

Lack of construct validity causes two deficits: [30]

- Contamination: When the scores obtained by the assessment tool represent features that are not part of the construct being studied.
- Deficiency: When there are aspects of the construct being studied that are not included in the assessment tool.

http://dx.doi.org/10.5772/67389

Some threats to construct validity are listed hereafter [2, 11, 13]:

- **Using non-reliable tools**: When reliability is threatened, the construct validity of the device can also be threatened.
- **Narrow stimulus sampling**: When the researcher studies a narrow sample or situation while the construct under study is much broader. Case studies are particularly subject to this threat.
- **Single operations:** When the construct is complex, but the measure inspects just one aspect of it. For example, the researcher measures only the "time spent with friends" as the indication of being happy. To avoid this threat, the researcher should make use of more indicators for the construct.
- **Single subject design:** This design is a threat to construct validity when it is used to implement an intervention or a treatment. This is because the individual's specifications may be responsible for the outcome resulted, not the intervention itself.
- **Experimenter expectancies**: It happens when a researcher has passion and some expectations about the outcome, in a way that it influences his/her interpretation and explanation of the results. This may cause an alternative explanation of the relationship between variables to be drawn, which decreases the construct validity, since it is not declaring the real circumstances of the construct under study.
- If by any means **participants get some clues about the study**, it will affect their performance in the experiment. This is because participant might presume the hypothesis or objective of the study, and act, in the sake of the objective, differently from their usual real behavior. This, in fact, changes the construct that has been assessed, since the study is assessing them when they are "motivated." To prevent or minimize this threat, researchers should attempt to provide fewer cues for participants.
- The same situation is established when participants from one group have **the tendency to compete** with the participants of the other group. This also makes them more motivated. To avoid this threat, researchers should minimize the incidental contacts between subjects.
- **Demoralization**: this happens when some individuals from one group are not satisfied with the treatment they are receiving, compared to the treatment that the other group receives. This makes them less motivated in participating and affects their performance. A solution for this threat is providing another valued treatment for the group that is suspected of being demoralized. Of course, this treatment should be a placebo or proved to have no intervening effect on the study.
- **Mono-method bias**: Happens when the researcher uses only one method of measurement. If this measurement has a poor construct or content validity, the study would be flawed. A method for minimizing this threat is using a number of methods at the same time, for example, questionnaire, self-report, and observation.
- **Mono-operation**: Happens when just one manipulation is used to affect the construct. For instance, to investigate the effect of a special drug as an intervention, participants are

divided to one placebo group and one treatment group. A more solid design will establish multigroup and sets different dosage of the drug for each group.

- **Poor construct definition:** This is when the construct is misdefined (e.g., assessing anxiety instead of depression), or has not been defined properly (e.g., assessing job satisfaction to represent overall happiness). Researchers should get advice from experts in the field before starting the study, to prevent this threat.

Figure 1 depicts a summary of this paper: the overall classification of validities and main issues that threaten different types of validity.

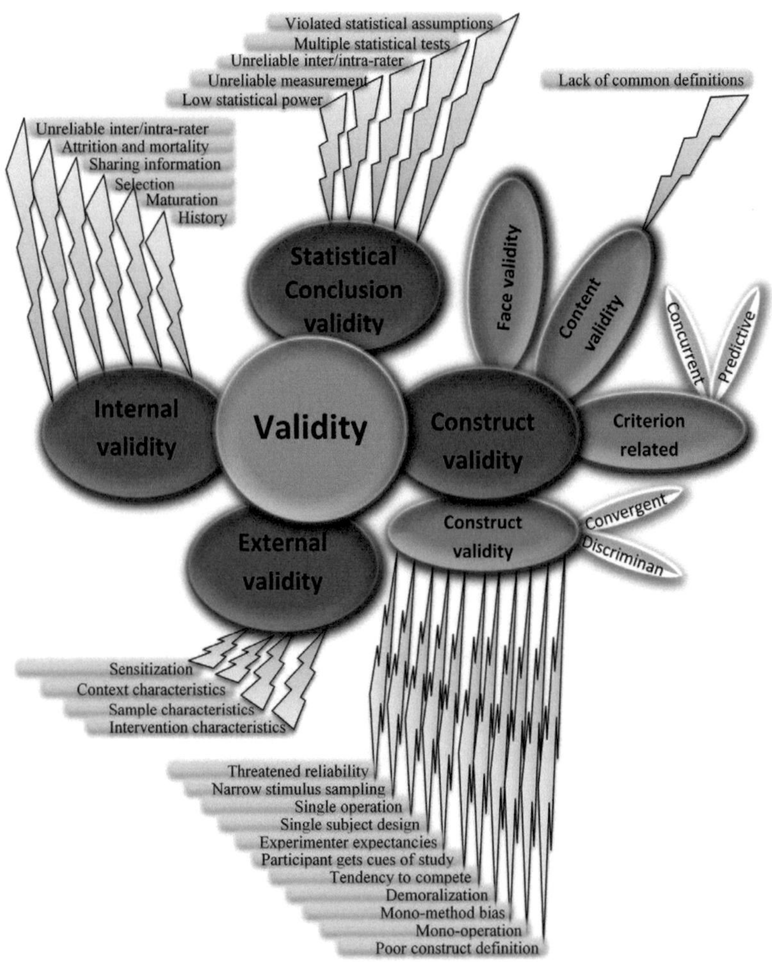

Figure 1. Overall classification of validities and main issues that threaten different types of validity (sparks).

http://dx.doi.org/10.5772/67389

4. Discussion

In this paper, four types of research validity with their subclasses were described. Now, one may ask, among these different types of validity, which one of them is more important and has priority in consideration in validity study? The answer is rather complicated due to different opinions existing in the literature, which will be described here.

Among different types of validity, construct validity is the one that more frequently has been subject to consideration, research, and publication (e.g., see [11, 13, 18]). This is because it appears that for some researchers, construct validity (and its subclasses) are the only concepts of research validity. Although this frequency of consideration relative to the other types of validity could be an indication of relatively greater importance for this type of validity, the authors could not find any explicit declaration of that.

However, as it was pointed out in the “statistical conclusion validity” section, Ottenbacher and Barrett Kathryn [1] believe that this type of validity (statistical conclusion validity) is the most important type of validity, though it has received little attention in rehabilitation research. The importance of this type of validity is that one should make sure that the findings are obtained as a result of a real covariation between variables, rather than chance.

Cook's and Compbell [28] and Mitchell [30], however, do not believe so, since they declare that the internal validity is the most important validity, and hence one should be more concerned about it. This is because this group of researchers believes that as long as a study does not have internal validity, the data achieved from it are not appropriate and trustworthy, and as a result they are not eligible to be generalized to other situations. Mitchell says that most authors are not concerned with external validity and they just put it under “further research” heading [30]. Bellini and Rumrill gather between these two former opinions for studies that investigate an unknown relationship between variables [2].

Some researchers, however, argue that at the end of the day, we need to generalize our data to other situations, and hence if they are valid but unable to be generalized, they are useless. They, therefore, believe that external validity deserves higher priority than what it has received in health research so far [34].

Another point that is worth noting here is the point drawn by Shadish et al. that construct validity is not a necessary condition for external validity, because “we can generalize across entities known to be confounded albeit less usefully than across accurately labeled entities” [35].

Bellini and Rumrill [2] state that in fact, all four research validities are important in turn, and one should try to have all of them in the higher level possible, but there is a logical order for them: first statistical conclusion validity should be established, to show that two variables covary. The second validity to be established is the internal validity, which focuses on the obtained relationship between variables. Then the researcher should be concerned with the construct validity which speaks about the construct that is involved in the relationship.

Eventually, the importance of external validity would arise that is concerned with the generalization of the results to other settings.

To sum up, it should be noted that all validities are of great importance of value for all research. However, one may not be able to maximize all of them at the same time, since maximizing one of them could be dependent on decreasing the other one. For instance, for increasing statistical conclusion validity with a given sample size, one may need to restrict the population under study, in order to decrease the variation between samples and increase the statistical power of the study. This, in turn, obviously leads to a reduction of generalizability of his/her results (lower external validity). Researchers, therefore, need to make trade-offs between different aspects of research validity—considering the conditions they are in, so that they could end up with the optimum validity for their research study.

Acknowledgements

The authors like to thank Dr. Susan Armijo-Olive for her invaluable guidance to writing this chapter. Also, we greatly appreciate the comments and feedback offered by Dr. Pamela Bentley.

Author details

Zohreh Salimi and Martin W. Ferguson-Pell*

*Address all correspondence to: martin.ferguson-pell@ualberta.ca

Faculty of Rehabilitation Medicine, University of Alberta, Edmonton, Alberta, Canada

References

[1] Ottenbacher KJ, Barrett Kathryn A. Statistical conclusion validity of rehabilitation research. Am J Phys Med Rehabil. 1991;**70**(1):138–143.

[2] Bellini JL, Rumrill PD. Perspectives on scientific inquiry: Validity in rehabilitation research. J Vocat Rehabil. 1999;**13**:131–138.

[3] Messick S. The interplay of evidence and consequences in the validation of performance assessments. Educ Res. 1994;**23**(2):13–23. DOI: 10.3102/0013189X023002013

[4] Messick S. Validity and washback in language testing. Lang Test. 1996;**13**(3):241–256. DOI: 10.1177/026553229601300302

[5] Messick S. Meaning and values in test validation: The science and of assessment. Educ Res. 1989;**18**(2):5–11. DOI: 10.3102/0013189X018002005

[6] Messick S. The standard problem meaning and values in measurement and evaluation. Am Psychol. 1975:955–966.

[7] Messick S. Standards of validity and the validity of standards in performance assessments. Am Educ Res Assoc. 1995;**14**(4):5–8.

[8] Messick S. Validity of physiological assessment: Validation of inferences from persons' responses and performances as scientific inquiry into score meaning. Am Psychol. 1995;**50**(9):741–749.

[9] Shepard LA. The centrality of test use and consequences for test validity. Educ Meas Issues Pract. 1997:5–13.

[10] Jonsson A, Svingby G. The use of scoring rubrics: Reliability, validity and educational consequences. Educ Res Rev. 2007;**2**(2):130–144. DOI: 10.1016/j.edurev.2007.05.002

[11] Brown JD. What is construct validity? Shiken: JALT testing and evaluation SIG newsletter. October 2000:8–12.

[12] Moss PA. Can there be validity without reliability?. Educ Res. 1994;**23**(2):5–12. DOI: 10.3102/0013189X023002005

[13] Shuttleworth M. Construct validity. Explorable. 2005;DOI: 10.4135/9781412950558.n102

[14] Cook DA, Beckman TJ. Current concepts in validity and reliability for psychometric instruments: theory and application. Am J Med. 2006;**119**(2):166.e7–166.e16. DOI: 10.1016/j.amjmed.2005.10.036

[15] McKay JR, Alterman AI, McLellan AT, Snider EC, O'Brein CP. Effects of random versus nonrandom assignment in a comparison of inpatient and day hospital rehabilitation for male alcoholics. J Consult Clin Psychol. 1995;**63**(1):70–78.

[16] Roberts BW, Kuncel NR, Shiner R, Caspi A, Goldberg LR. The power of personality: the comparative validity of personality traits, socioeconomic status, and cognitive ability for predicting important life outcomes. Perspect Psychol Sci. 2007;**2**(4):313–375. DOI: 10.1111/j.1745-6916.2007.00047.x

[17] Brennan L, Voros J, Brady E. Paradigms at play and implications for validity in social marketing research. J Soc Mark. 2011;**1**(2):100–119. DOI: 10.1108/20426761111141869

[18] Arthur W, Woehr DJ, Maldegen R. Convergent and discriminant validity of assessment center dimensions: A conceptual and empirical reexamination of the assessment center construct-related validity paradox. J Manage. 2000;**26**(4):913–835. DOI: 10.1177/014920630002600410

[19] Hunter JE, Schmidt FL, Hunter R. Differential validity of employment tests by race: A comprehensive review and analysis. Psychol Bull. 1979;**86**(4):721–735. DOI: 10.1037//0033-2909.86.4.721

[20] Lynn MR. Determination and quantification of content validity. Nurs Res. 1985;**35**(6): 382–386.

[21] Longshore D, Stein JA. Self-control in a criminal sample: An examination of construct validity. Criminology. 1996;**34**(2):209–228.

[22] Hoffmann A, Skalsky JR. Inferential properties of an individual-based survival model using release-recapture data: sample size, validity and power. J Appl Stat. 1995;**22**(5&6):579–595.

[23] Dunbar CC, Robertson RJ, Baun R, et al. The validity of regulating exercise intensity by ratings of perceived exertion. Med Sci Sports Exerc. 1992;**24**(1):94–99.

[24] Stunkard AJ, Messick S. The three-factor eating questionnaire to measure dietary restraint, disinhibition and hunger. J Psychomatic Res. 1985;**29**(I):71–83.

[25] Coste J, Judet O, Barre O, Siaud JR, Cohen de Lara A, Paolaggi JB. Inter- and intraobserver variability in the interpretation of computed tomography of the lumbar spine. J Clin Epidemiol. 1993;**47**(4):375–381.

[26] Banks JL, Marotta CA. Outcomes validity and reliability of the modified Rankin scale: implications for stroke clinical trials: a literature review and synthesis. Stroke. 2007;**38**(3):1091–1096. DOI: 10.1161/01.STR.0000258355.23810.c6

[27] Faes MC, Reelick MF, Perry M, Olde Rikkert MGM, Borm GF. Studies with group treatments required special power calculations, allocation methods, and statistical analyses. J Clin Epidemiol. 2012;**65**(2):138–146. DOI: 10.1016/j.jclinepi.2011.05.007

[28] Cook TD, Compbell DT. Quasi-Experimentation: Design & Analysis Issues for Field Settings. Chicago: Rand McNally; 1979.

[29] Campbell DT, Stanley JC. Experimental and Quasi-Experimental Designs for Research. Boston: Houghton Mifflin Company; 1963.

[30] Mitchell TR. An evaluation of the validity of correlational research conducted in organizations. Acad Manag Rev. 1985;**10**(2):192–205.

[31] Haynes SN, Richard DCS, Kubany ES. Content validity in psychological assessment: a functional approach to concepts and methods. Am Psychol Assoc. 1995;**7**(3):238–247.

[32] Kaplan RM, Bush JW, Berry CC. Health status: types of validity and the index of well-being. Health Serv Res. 1976;**11**(4):478–507.

[33] Peter JP. Construct validity: a review of basic issues and marketing practices. J Mark Res. 1981;**18**(2):133–145.

[34] Steckler A, McLeroy KR. The importance of external validity [Editorial]. Am J Public Health. 2008;**98**(1):9–10. DOI: 10.2105/AJPH.2007.126847

[35] Shadish WR, Cook TD, Campbell DT. Experimental and Quasi-Experimental Designs for General Causal Inference. Boston: Houghton Mifflin Company; 2002.

Chapter 2

Rare and Disabling Movement Disorders: An Indian Experience

Gurusidheshwar Wali and Gautam Wali

Additional information is available at the end of the chapter

http://dx.doi.org/10.5772/67598

Abstract

Recent decades have seen exciting developments in the field of movement disorders. These include identification of rare clinical syndromes and use of technological advances to understand their pathogenesis. Three such disorders are discussed here. The description of the Uner Tan syndrome from Turkey and surrounding regions provoked research into the controversial field of genetically induced devolution. Such cases with few additional findings have now been described from India. Sepiapterin reductase deficiency is a rare treatable autosomal recessive form of dopa responsive dystonia. Indian literature has recently added five confirmed cases to the international database. Such cases are eminently treatable. Successful application of modern technology in understanding the pathogenesis of progressive neurodegenerative disorder has been highlighted in the section on hereditary spastic paraplegias. Hitherto undescribed subcellular organelle transport defects and their potential rectification with known drugs have been demonstrated raising hopes for their cure.

Keywords: movement disorders, Uner Tan syndrome, neurogenetics, sepiapterin reductase deficiency, hereditary spastic paraplegia, stem cell modelling of HSP

1. Introduction

Movement disorders is a relatively young but fast growing sub-speciality of neurology. Recent developments in the clinical, genetics and treatment aspects of this group of disorders are exciting. The nature of movement disorders seen in India differ from the western cases. The vast population of the country where consanguineous marriages are still prevalent is a source of large number of rare movement disorders. Modern technology including whole genome sequencing has promoted in-depth understanding of such cases. In addition, such rare disorders can lead to physical disability especially in children when they remain undiagnosed and untreated. This paper highlights selected Indian contributions to such developments with an intention to increase awareness about such rare disorders. Some of them are potentially treatable. The topics discussed here include Uner Tan syndrome, sepiapterin reductase deficiency and hereditary spastic paraplegia.

2. Uner Tan syndrome

In the year 2005, Uner Tan from Turkey discovered and described an autosomal recessive genetic syndrome, at present, recognized as Uner Tan syndrome (UTS), which composed of habitual quadrupedalism, impaired intelligence, primitive language and no conscious experience [1]. The cases described by him had quadrupedal locomotion (Q/L) using two palms and two feet with extended legs. Some of them later developed bipedal locomotion (B/L) although Q/L remained the preferred mode of locomotion. Recently he has described similar cases presenting with infantile hypotonia which disappeared by adolescence to be replaced by quadrupedal locomotion [2]. The cases without infantile hypotonia are now identified as UTS Type I and those associated with infantile hypotonia are identified as UTS Type II cases. Considering the unusual psychomotor behaviour of these cases, resembling that of sub-human primates, Uner Tan proposed the theory of 'genetically induced devolution' to explain the genesis of this syndrome. This theory has been questioned by other researchers [3]. This chapter documents two new families of UTS from India. Additional features of dystonia, which appear with their bipedal locomotion, are highlighted.

2.1. Indian cases of UTS

The present chapter describes clinical and radiological findings in three cases of UTS belonging to two Indian families [4] (**Figure 1**). Both of these families resided in Mantur, a small village belonging to Belagavi district in the state of Karnataka. The village is close to semi urban cities with exposure to modern lifestyle practices. Family A had one female case, whereas Family B had two male cases affected by the syndrome. Autosomal recessive type of transmission was identified in each of the families. There was no history of birth asphyxia, trauma or encephalitis in any of the cases. Haematological and biochemical tests, including serum parathormone, vitamin D3, B-12, copper and ceruloplasmin levels, were measured and found to be within normal limits. The VLDL receptor gene was negative in all the cases. For sake of brevity, the girl belonging to Family A is designated as Case A-1 and the brothers belonging to Family B are designated as Case B-1 (elder) and Case B-2 (younger), respectively.

2.1.1. Case A-1

This girl who was born of post-term pregnancy as a large baby had remained floppy until the age of 5 years. She developed Q/L at the age of 6 years and B/L at the age of 11 years. She had features of severe mental retardation with a MMSE score of 0/30. Speech was absent but she repeatedly mumbled a meaningless word. She responded to few commonly used non-verbal commands. Self-care was poor. Most often she moved using Q/L, but at times attempted walking with a wide-based ataxic bipedal gait. She suffered sleep onset epileptic seizures

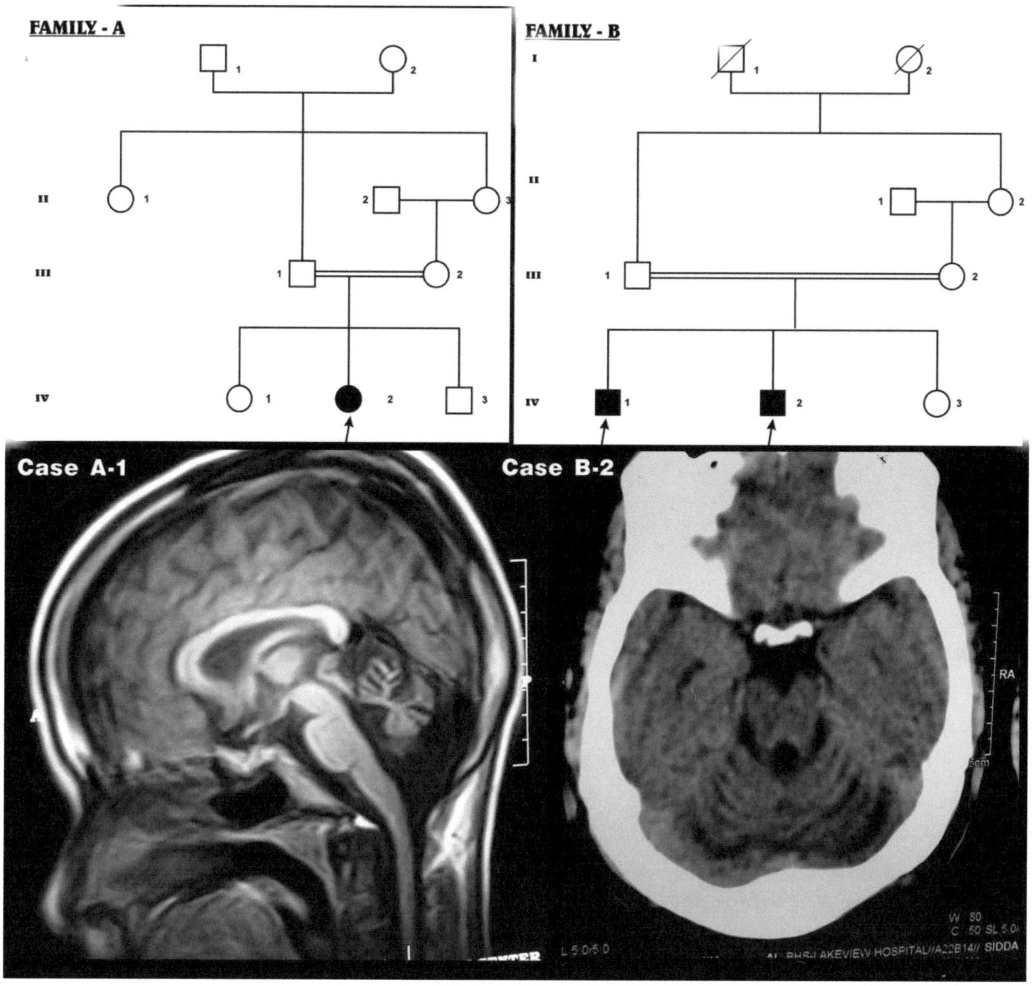

Figure 1. The top section of the figure shows the genogram of Families A and B. The lower section of the figure shows MRI/CT brain images of patients A-1 and B-2, respectively. The images reveal the presence of cerebellar atrophy.

from the age of 18 months, which were partly controlled with phenobarbitone. MRI of the brain revealed moderately severe hypoplasia of the cerebellum including the vermis along with mild cerebral cortical thinning.

2.1.2. Case B-1

This 35-year-old male who was born of full-term normal home delivery had no infantile hypotonia. Features of severe mental retardation were present with no speech but he was able to follow few non-verbal commands. MMSE score was 0/30 and self-care was poor. He exhibited severe autistic behaviour avoiding eye to eye contact with others. He used only Q/L which had appeared at the age of 7 years (**Figure 2**). He never attempted B/L. He could be made to stand with support for few seconds using a wide base support. He suffered rare episodes of generalized tonic-clonic seizures which remained untreated. Neuroimaging was not possible in him.

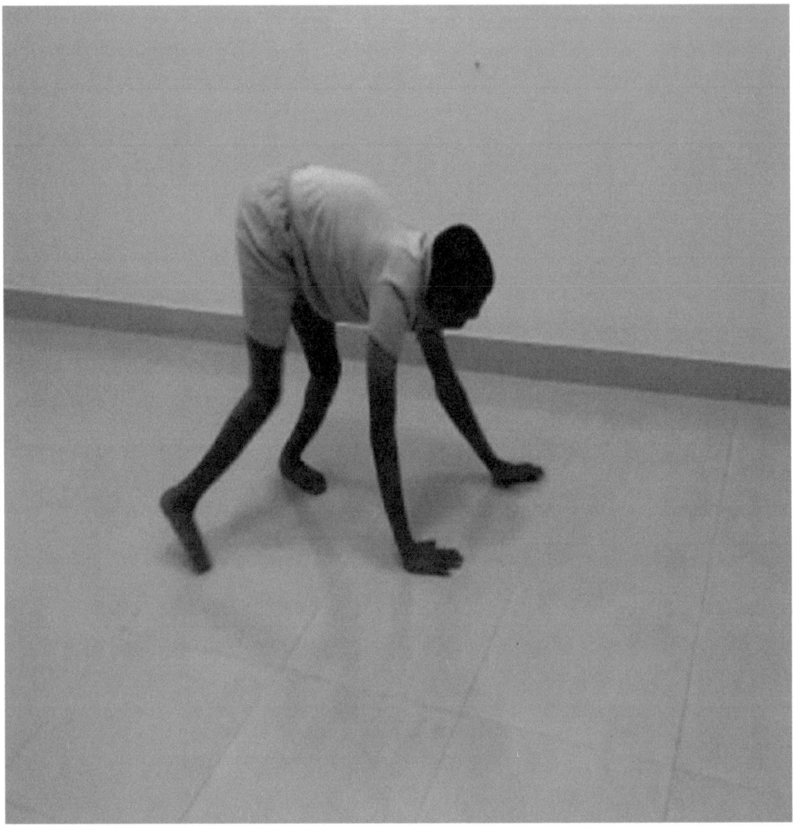

Figure 2. The figure is a still photography of Case B-1 showing the quadrupedal locomotion. Note the diagonal sequencing of the limbs and the extension of the lower extremities at the knee.

2.1.3. Case B-2

This 30-year-old younger brother of Case B-1 and born of normal home delivery had evidence of severe mental retardation with no speech and MMSE score of 0/30. He exhibited autistic behaviour shunning crowds and often strayed away silently. He had developed Q/L at the age of 7 years and B/L at the age of 25 years. His bipedal gait was wide-based and short-stride but not ataxic. He had camptocormia and almost continuous dystonia abductor posturing of his right arm during bipedal locomotion. Presently, he prefers to move bipedally for most of the time. A CT scan of the head revealed evidence of cerebellar hypoplasia.

2.1.4. Discussion

The diagnosis of UTS was made in these cases based on the presence of the cardinal features of the syndrome as mentioned above.

Most cases reported by Uner Tan resided in remote villages of Turkey having poor communication with the outer world. He proposed that their alienation from outside world,

http://dx.doi.org/10.5772/67598

low socio-economic status, poor nutrition and poor parental care may be contributing to the causation of the syndrome [1]. In contrast although our cases resided in a village, they were exposed to modern lifestyle practices in view of close proximity to urban settings. These sibs lived protected as members of 'nuclear families', a concept still prevalent in rural India.

In addition to the cardinal features of the syndrome, these patients had (i) microcephaly with facial dysmorphism, (ii) restrictive repetitive behaviours and stereotypies and (iii) poor social interaction. Head circumference in relation to body height measurements mentioned in the table indicates the presence of mild microcephaly in all the cases. While Case A-1 had the feature of repetitive mumbling of a meaningless word, a form of verbal stereotypy, Case B-1 and Case B-2 had extremely shy behaviour avoiding eye to eye contact, social interaction and communication. Dystonic limb posturing was present in cases 2 and 3 during attempted bipedal locomotion.

The pathophysiology of habitual quadrupedalism is unclear. The role of central pattern generators (CPG) present in the cervical and the lumbosacral spinal segments has been discussed in this regard. Reorganization and adaptation of CPG networks is possibly involved in the appearance of human quadrupedal locomotion [5]. Considering the complex balance mechanisms needed for a well-balanced bipedal locomotion, walking on all the four is much easier since the overall base of support of the body is better in the latter type of locomotion [6]. In this regard, the possibility of quadrupedalism appearing as an epiphenomenon secondary to very early onset balance problem cannot be ruled out. The Indian cases were noted to have mild forms of limb dystonia, which became obvious when they attempted bipedal locomotion. In one of his reviews, Uner Tan pointed out that basal ganglia signs were absent [7]. Future studies in UTS may need to incorporate these clinical features to revise our present understanding of this interesting syndrome.

In summary, the clinical observations made in the Indian cases of UTS suggest that components of limb dystonia may be part of the clinical presentation of the syndrome.

3. Sepiapterin reductase deficiency, a rare but treatable paediatric movement disorder

Dopa responsive dystonias (DRDs) are a genetically heterogeneous subgroup of paediatric neurotransmitter diseases (PNDs), which occur due to defective tetrahydrobiopterin (BH4) metabolism. Since hyperphenylalaninemia, which is a useful marker in the screening of such diseases, is not always present, the disorders are classified based on the presence or absence of hyperphenylalaninemia. BH4 defects without hyperphenylalaninemia include dopa responsive dystonia (Segawa disease), sepiapterin reductase deficiency (SRD) and dihydrobiopterin reductase deficiency. BH4 defects without hyperphenylalaninemia include autosomal recessive GTP-1 cytohydroxylase deficiency, 6-pyruvoyl-teytrahydropterin synthase deficiency (6-PTS) and dihydropteridine reductase deficiency without hyperphenylalaninemia. In addition primary defects of monoamine synthesis include Tyrosine hydroxylase deficiency (TH) and Aromatic L-amino acid decarboxylase deficiency (AADC).

Examples of autosomal dominant and recessive types of dopa responsive dystonias are described. Segawa disease with autosomal dominant inheritance is a landmark in this group of diseases. Sepiapterin reductase deficiency (SRD) is a rare example of autosomal recessive variety [8]. It is caused by mutation in the SPR gene, located on the chromosome 2p14-p12 [9]. The clinical presentation includes cognitive delay along with dopa responsive mixed motor disorder often masquerading as cerebral palsy [10]. Neonatal screening tests are not helpful in the diagnosis of this disease since it is not associated with hyperphenylalaninemia. However the presence of certain additional clinical features like oculogyric crisis, diurnal fluctuation of the symptoms and hypersomnia provide a clue to the clinical suspicion. Clinical response to dopamine therapy is often rewarding and at times dramatic. The laboratory diagnosis of SRD is based on CSF analysis of neurotransmitter metabolites and pterins. The classical CSF findings in SRD include increased levels of sepiapterin along with reduced levels of 5-hydroxyindolacetic acid and homovanillic acid. The diagnosis can be confirmed by studying the SR activity in the fibroblasts and mutational analysis of the SPR gene.

The following paragraphs provide an overview of the clinical profile of SRD based on the information obtained from various case reports and recent reviews [11, 12] of Indian cases have been summarized.

3.1. Demography

Till date, a total number of 43 cases have been recognized from all over the world and are reviewed by Friedman et al. [11]. These include cases which are published and those documented in the international database maintained by the PND society. Both genders are equally represented. The cases belong to a wide ethnic background from various countries including India. A founder effect was suspected in the seven patients reported from Malta.

3.2. Clinical profile

3.2.1. Mixed motor signs and symptom

SRD causes symptoms related to dopamine and serotonin metabolism disturbances. The salient clinical features include motor and language delay associated with axial hypotonia, limb weakness, dystonia with diurnal fluctuation and oculogyric crisis. The latter group of symptoms often show sleep benefit. Other symptoms include parkinsonism, limb hypotonia, hypertonia and dysarthria. As against Segawa disease, the occurrence of dystonia is not universal in SRD. It is present in only about half of the cases. Less common neurological features include chorea, myoclonus and seizures.

The symptoms of SRD usually appear in age-related pattern [12]. The symptoms of axial hypotonia, tremors, hypersomnia, oculogyric crisis and sleep benefit usually appear in the first year of life while limb spasticity, diurnal fluctuation, dysarthria and dystonia become obvious between the ages of 2 and 6 years. The triad of oculogyric crisis, paroxysmal stiffening and axial hypotonia may be considered as a strong clue to suspect the diagnosis of SRD in the infantile group. However, these symptoms may persist or appear during late childhood as well. It is not unusual for the paroxysmal episodes to be misdiagnosed as seizure

http://dx.doi.org/10.5772/67598

phenomenon resulting in delayed diagnosis and unwanted use of anticonvulsants. 'Cerebral palsy' is the commonest diagnosis in majority of the cases before reaching a final diagnosis.

3.2.2. Non-motor neurological features

These include sleep disturbances, behavioural changes and cognitive disabilities. Hypersomnia is the commonest sleep disturbance and present in nearly 50% cases especially in the infantile age group. Behavioural changes are also seen in slightly more than half of the cases. Cognitive disabilities of varying degree are seen in almost 90% cases. Nearly half of them have mental retardation, whereas others have mild-to-moderate disability.

3.2.3. Other features

Non-specific non-neurological features have been reported in SRD. These include pulmonary symptoms, vegetative symptoms such as gastrointestinal disturbances, hyperhidrosis, premature greying of hairs and menstrual abnormalities.

3.2.4. Salient investigations

CSF analysis is the mainstay of the diagnosis of SRD. The classical findings include

i) Low 5-hydroxyindolacetic acid and homovanillic acid levels.

ii) Elevated total biopterin, dihydrobiopterin and sepiapterin.

The study of SR activity in the fibroblast cultures has been performed in few cases and found to be confirmative. SR activity is either less or absent in such cases.

Genotyping of the documented 42 cases has revealed 16 mutations in the SPR gene. The intronic homozygous variant c596-2A>G is the most common genotype.

3.2.5. Treatment responses and outcomes

Substitution therapy is the mainstay of treatment in SRD. Most cases were treated either with L-dopa/carbidopa or benserazide (doses ranging from 1.45 mg to 20 mgm/kg/day) or 5-hydroxytryptophan (doses ranging from 0.65 to 5.9 mgm/kg/day). Combination therapy is also useful and in some cases better than the use of single drug. Other drugs used to treat include BH4, bromocriptine, anticholinergics, selegiline, sertraline and melatonin.

The response to treatment is always rewarding and at times dramatic. Patients may show improvement in motor activity and sleep disturbances within few hours of administering medication. Children who have been chair-bound or bed-bound may start moving and enjoy the unexpected relief with tremendous excitement. The rapid recovery is usually associated with appearance of choreiform dyskinesias requiring dose titration. Few cases show partial improvement. Favourable response to treatment is noted even in those cases who have received the treatment late in the course of their illness due to delayed diagnosis. Long-term follow-up has shown that few cases may require mild escalation in the dose of the drugs to maintain the favourable benefit. It also shows that response to therapy is not as complete as seen in Segawa disease. Residual motor and cognitive dysfunction may persist. Early treatment is associated with better recovery.

3.2.6. Indian cases of SRD

Five Indian cases (three males and two females) of chemically and genetically proven SRD have been documented in the international database maintained by the PND society (BIODEF database IDs # 526,527,637,638 and 639) [13]. All of these cases have been reported from the Belgaum region of North Karnataka [11, 14]. These cases with consanguineous parenthood belonged to two independent families. The referring diagnosis in all these cases was 'cerebral palsy'. All cases showed very favourable and sustained response to L-dopa therapy (**Figure 3**). They have been followed up clinically for the past 5 years. The average maintenance dose of L-dopa was 1.5–2 mg/kg body weight in all these cases. All cases had the *core symptom complex* inclusive of psychomotor delay, axial hypotonia, limb hypotonia/hypertonia, acral dystonia/athetosis, diurnal variation and sleep benefit. **Table 1** summarises the clinical data which was present in addition to this core symptom complex.

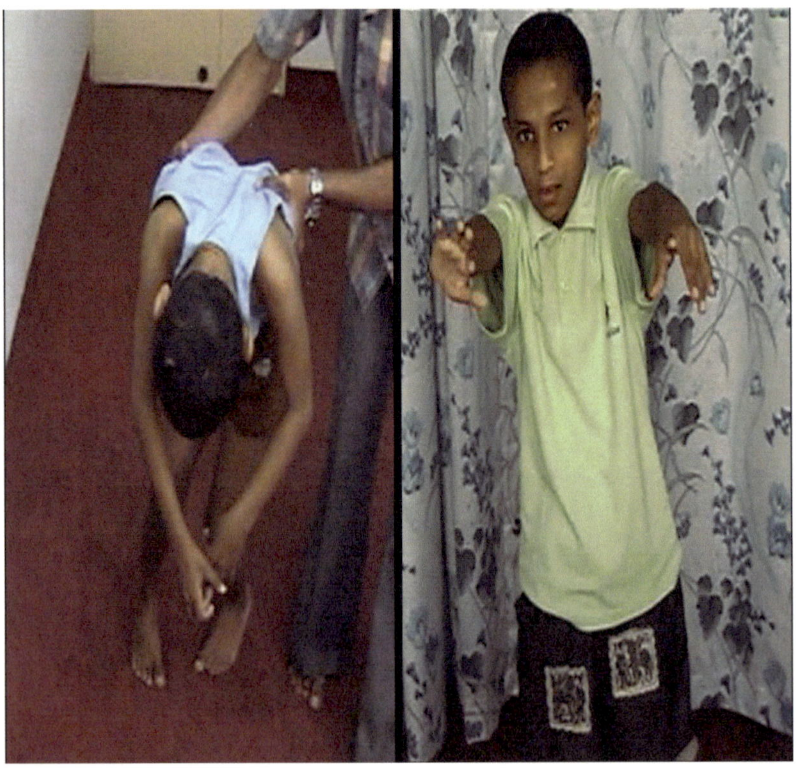

Figure 3. This figure reveals the clinical state of the patient (BIODEF #523) of SRD before and after administration of levodopa. The left half of the figure shows evidence of truncal hypotonia and spastic-dystonia limb posturing. The right half of the figure shows the clinical state of the patient after administration of a single low dose of levodopa/carbidopa. The response was dramatic and seen within 12 hours.

http://dx.doi.org/10.5772/67598

Case no. sex	Onset age	Age at diagnosis	Features additional to core symptom complex	Response to L-dopa/carbidopa therapy	5-year follow-up
1. Male	3 months	10 years	Premature greying of hairs/wheezing	Dramatic and well sustained to date	Independent, attends school and plays football
2. Female	3 months	11 months	Wheezing	Dramatic and well sustained to date	Independent, attends school and games
3. Male	9 months	5 years	Two episodes of seizures	Improvement noted after a week of initiating L-dopa therapy. Well-sustained to date	Independent. Attends school and regular games
4. Male	3 months	3 years	Irritable behaviour/ bilateral talipes equinovarus	Improvement noted after 1 week of initiating L-dopa therapy. Well-sustained to date	Independent and attends school. Scholastic performance is average
5. Female	3 months	1 year	Irritable behaviour with incessant crying/bilateral talipes equinovarus	Improvement noted after 1 week of initiating L-dopa therapy. Well-sustained to date	Stands and walks few steps independently. Does not attend school. Irritable behaviour

Table 1. Clinical summary of Indian cases of SRD.

It is likely that SRD is more common all over the world but possibly remains under-diagnosed due to lack of awareness and inadequate laboratory services. Such cases are likely to masquerade as 'cerebral palsy' and remain untreated for life.

4. Hereditary spastic paraplegia

Hereditary spastic paraplegia (HSP) is an untreatable rare genetic disorder, which causes long-term physical disability in the affected individuals. There is a high degree of genetic heterogeneity with over 60 causative genes identified so far. A collaborative study using whole genome sequencing (WSG) was performed in 10 HSP families from Belgaum in Karnataka, India. Pathogenic variants were identified in four consanguineous families. This included a novel frameshift homozygous deletion in *CYP2U1* (c.782_785delTCTG) gene in one family and a novel homozygous donor splice site variant in *DDHD2* (c.1125+1G>T) gene in another family. A probable genetic cause was identified in 40% of the families limited to consanguineous families in whom a family study was possible. Mutations were found in genes causing metabolic disorders such as Zellweger spectrum disorder and GM1 gangliosidosis. This study supports a role of WGS as a diagnostic tool in HSP [15].

4.1. Stem cell modelling of HSP

HSP is characterised by degeneration of long axons along the corticospinal tract, leading to lower limb spasticity. The mechanisms underlying HSP mutations that lead to degeneration of the long axons are unclear. Researchers have recently used patient-derived cells from the olfactory mucosa, a population of neural progenitor cells, derived from biopsies of the olfactory mucosa from HSP patients with SPAST mutations and from healthy controls, in order to identify cell functions altered in HSP [16]. Mutations in the SPAST gene account for the largest group of adult-onset HSP patients. These researchers show that SPAST patient-derived cells have reduced expression of protein Spastin (encoded by SPAST), reduced expression of stabilized microtubules, alterations in transport of cellular cargo and increased oxidative stress [17] (**Figure 4**). Based on their findings, Gautam Wali and colleagues propose that the downstream effects of SPAST mutations may cause a chronic state of oxidative stress in cortical motor neurons and other neurons, which ultimately leads to their degeneration, eventually manifesting the disease. They also show that the defects observed in the patient cells can be rescued using a microtubule-binding drug, Epothilone D. Epothilone D can cross the blood-brain barrier, making it a potential drug for HSP therapy [18].

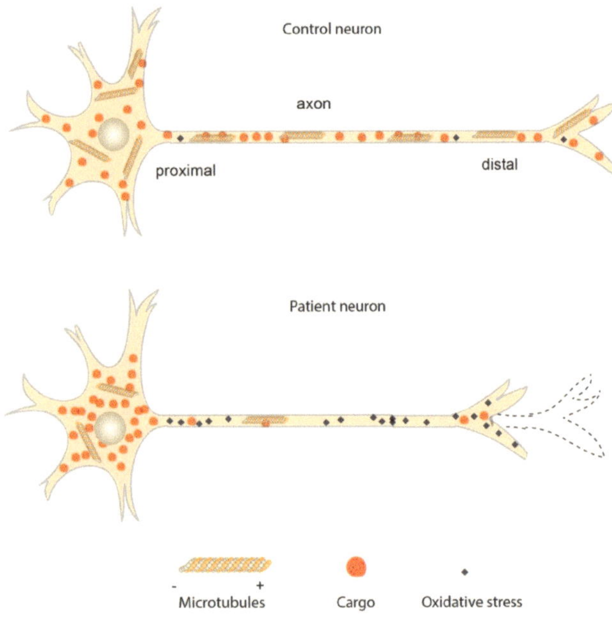

Figure 4. This figure summarises the mechanism of cell organelle dysfunction in HSP as confirmed by Gautam Wali and colleagues using patient-derived olfactory stem cells. They suggest a mechanism whereby *SPAST* mutations lead to reduced levels of stable microtubules which compromises cargo trafficking and leads to increased oxidative stress. These downstream effects of SPAST mutations may cause a chronic state of oxidative stress in cortical motor neurons and other neurons, which ultimately leads to their degeneration, eventually manifesting the disease. They also showed that Epothilone D, a microtubule-binding drug, can rescue the cell function defects in patient cells, making it a potential candidate for a future therapy for HSP.

Author details

Gurusidheshwar Wali[1,2]* and Gautam Wali[3]

*Address all correspondence to: walidoc@hotmail.com

1 K.L.E.University, Belgaum, Karnataka, India

2 Neurospecialities Centre, Belagavi, India

3 Department of Neurogenetics, Kolling Institute of Medical Research, Sydney

References

[1] Tan U. Uner Tan syndrome: quadrupedality, primitive language, and severe mental retardation: a new theory on the evolution of human mind. Neuro Quantol 2005;4:250–255.

[2] Tan U. Two families with quadrupedalism, mental retardation, no speech and infantile hypotonia (Uner Tan Syndrome Type II); a novel theory for the evolutionary emergence of human bipedalism. Front Neurosci 2014;8:84.

[3] Humphrey N, Mundlos S, Turkmen S. Genes and quadrupedal locomotion in humans. Proc Natl Acad Sci USA 2008;105:E26.

[4] Wali G. Two Indian families with quadrupedal locomotion resembling Uner Tan syndrome: a video document. Mov Dis Clin Pract 2016;4:49–50. DOI: 10:1002/mdc3.12380

[5] MacLellan MJ, Ivanenko YP, Cappellini G, Sylos Labini F, Lacquaniti. Features of hand-foot crawling behaviour in human adults. J Neurophysiol 2012;107:114–125.

[6] Lacquaniti F, Grasso B, Zago M. Motor patterns in walking. News Physiol Sci 1999;14: 168–174.

[7] Tan U. Uner Tan syndrome: history, clinical evaluations, genetics and the dynamics of human quadrupedalism. The Open Neurol J 2010;4:78–89.

[8] Blau N, Bonafe L, Thony B. Tetrahydrobopterin deficiencies without hyperphenylalaninemia: diagnosis and genetics of dopa responsive dystonia and sepiapterin reductase deficiency. Mol Genet. Metab 2001;74:172–185.

[9] Bonafe L, Thony B, Penzien JM, et al. Mutations in the sepiapterin reductase gene cause a novel tetrahydrobiopterin dependent monoamine neurotransmitter deficiency without hyperphenylalaninemia. Am J Hum Genet. 2001;69:269–277.

[10] Neville BGR, Parascandalo R, Farrugia R, Felice A. Sepiapterin reductase deficiency: a congenital dopa-responsive motor and cognitive disorder. Brain 2005;128:2291–2296.

[11] Friedman J, Roze E, Abdanur JE, Chang R et al. Sepiapter in reductase deficiency: a treatable mimic of cerebral palsy. Ann Neurol 2012;71:520–530.

[12] Dill P, Wagner M, Somerville A, Thony B, Blau N Weber P. Paroxysmal stiffening, upward gaze and hypotonia. Hallmarks of sepiapterin reductase deficiency. Neurology 2012;78: e 29–32.

[13] www.biopku.org

[14] Wali GM, Thony B, Blau N. Sepiapterin reductase deficiency : two Indian siblings with unusual clinical features. Mov Disord 2010;25:954–955.

[15] Kishore R Kumar, Gurusidheshwar Wali, Mahesh Kamate, et al. Defining genetic causes of hereditary spastic paraplegia with whole genome sequencing. 13th International Congress of Human Genetics, Kyoto, Japan 2016; Abstract 91220, 2014;42(3):174183.

[16] Abrahamsen G, Fan Y, Matigian N, et al. A patient-derived stem cell model of hereditary spastic paraplegia with SPAST mutations. Dis Models Mech 2013;6(2):489–502.

[17] Wali G, Sutharsan R, Fan Y, et al. Mechanism of impaired microtubule-dependent peroxisome trafficking and oxidative stress in SPAST-mutated cells from patients with Hereditary Spastic Paraplegia. Sci Rep 2016;6:27004.

[18] Fan Y, Wali G, Sutharsan R, et al. Low dose tubulin-binding drugs rescue peroxisome trafficking deficit in patient-derived stem cells in Hereditary Spastic Paraplegia. Biol Open 2014;3:494–502.

Chapter 3

Five-Wheeled Wheelchair with an Add-On Mechanism and Its Semiautomatic Step-Climbing and -Descending Function

Masayoshi Wada and Yu Munakata

Additional information is available at the end of the chapter

http://dx.doi.org/10.5772/67558

Abstract

In this chapter, we propose a novel add-on electric drive system for propelling a manual wheelchair on the floor together with an advanced function to climb and descend a step with no human support. The proposed add-on mechanism consists of an active-caster drive wheel and a reconfigurable link mechanism with a linear actuator to change the location of the drive wheel relative to a wheelchair. By attaching the mechanism to a manual wheelchair, we build a five-wheeled wheelchair. Since the drive wheel is attached on the back of the wheelchair, a risk of falling to the back is significantly reduced. To surmount a step with no help, we develop a step-climbing and -descending strategy by using the proposed wheelchair with a reconfigurable link mechanism. The five-wheel configuration guarantees a static stability of the wheelchair when some wheels are hovered from the ground. The function is used in step-climbing and -descending strategies to realize the transfer of a wheelchair user. In order to reduce the effort of a wheelchair user to control the complicated step surmount strategies, semiautomatic system is installed on the prototype wheelchair whose availability is verified through experiments.

Keywords: active-caster, wheelchair add-on electric drive, step-climbing and -descending, semiautomatic control

1. Introduction

For the aging society worldwide, need for machines that can support human moving capabilities gets more and more importance in recent years. Electric wheelchairs [1] are one of the most popular solutions for overcoming the aging problem in the past few decades. Electric-powered wheelchairs are used as one of the moving support devices not only for injured people but also for the elderly. Among the electric-driven wheelchairs, a "motorized wheelchair" has become popular in recent years. Different from a standard electric wheelchair, which is designed as electric-driven, a motorized wheelchair is composed of a manual wheelchair frame and an add-on electric drive module. Because it is foldable and light weight, users may carry it by cars to travel long distances easily.

A pair of electric drive modules for propelling left and right large wheels of a manual wheelchair are popular and commercially available (**Figure 1(a)**) [2]. Since the main parts are from a manual wheelchair, the diameter of a front caster is quite small, which prevents a motorized wheelchair from passing over a few centimeters step. Addition to this, if drive motors on the left and right provides large torque when the front casters stack into the gap or step, a wheelchair might fall in back since the reaction torques of the drive motors are applied to the wheelchair frame. In contrast, a five-wheeled wheelchair composed of a manual wheelchair and a single-wheel drive system was proposed (**Figure 1(b)**) [3]. The installed drive wheel is a standard orientable wheel, which can change its orientation at a steady point on the ground. This type of wheel has a nonholonomic constraint in its sideways direction and cannot control two degrees of freedom (2DOF) motion independently.

Toward the problem of step-climbing and -descending of wheelchairs, a lot of systems and mechanisms have been conducted. Wheelchairs with additional legs [4], some spoke wheels [5, 6], step-climbing method by the dynamic control as the inverted pendulum [7, 8], triangle frame with three wheels at the triangle vertices [9–11], and tracks on each side [12] have been proposed. The iBot wheelchair [13], which has the dynamic balancing functionality as an inverted pendulum, and Scaleve [14], which has additional tracks to climb over steps, have also been introduced. However, they are too heavy to carry them by personal cars.

A step-climbing method of the front casters with steering shafts fixed by the special mechanism [15], and a step-climbing and -descending strategies with a partner robot [16] are proposed to enhance the step-climbing capability of a manual wheelchair. The former is the step-climbing method only for the front casters and step-descending method has not been reported. The real-world applications of the latter system are quite limited.

In this paper, we propose an add-on mechanism to satisfy the step-climbing and -descending strategies, and in the both the strategies, the wheelchair approaches a step from the front. In our previous works, we proposed a five-wheeled wheelchair equipped with an add-on mechanism on the back of a manual wheelchair [17]. The add-on mechanism is composed of an active-caster, a linear actuator, and some link frames, which is connected to a manual

(a) Dual-wheel drive system[2] (b) Single-wheel drive system[3]

Figure 1. Conventional motorized wheelchair. (a) Dual-wheel drive system [2]. (b) Single-wheel drive system [3].

wheelchair frame. The active-caster [18] is used as a propelling mechanism for the manual wheelchair. The linear actuator is used as the wheelchair's posture change mechanism. In the climbing process of the front casters, they can be hovered from the ground by changing the location of the active-caster to backward of the wheelchair using the linear actuator. The large wheels and the drive wheel can be hovered from the ground, and pass over a step using the same method as that in the front casters. In the descending motion, the large wheels of the wheelchair can be landed on the ground by applying the hovering motion of the front casters to avoid the wheelchair frame from leaning forward. We derive the mechanical design condition for satisfying the static stability during the proposed step-climbing and -descending strategy. We determine the link configuration parameters of the add-on mechanism based on the derived mechanical condition and that of the step-climbing strategy. The descending abilities are realized together with the step-climbing function by constructing the add-on mechanism prototype. Also, we propose the semiautomatic control system for the proposed step-climbing and -descending strategies. By using the step detecting system, the front casters are hovered from the ground. Moreover, we propose the control method of the add-on mechanism in each process of the step-climbing and -descending motion. We have subsequently confirmed that the user could pass over a step automatically, using the proposed step-climbing and -descending strategies after the prototype assembly.

2. An add-on drive system

2.1. Active-caster

Figure 2 shows a schematic top view of an active-caster and an overview of the drive unit whose wheel axis is located in an offset position of the steering axis as a passive caster. The distance between the axes is called "caster-offset" as indicated by a parameter "s" in the figure.

The active-caster is used as a drive wheel by actuating a wheel shaft and a steering shaft by respective motors. To provide a traction power in an arbitrary direction with an arbitrary magnitude, a wheel rotation and a steering rotation are coordinated by a control law represented in Eq. (1). The equation includes cosine and sine functions of wheel orientation θ to derive the wheel and steering rotations, ω_w and ω_s. The wheel angle θ is measured by an absolute angle sensor and drive wheel, and steering axes are driven by the independent motors in appropriate ratios of angle θ, as represented in Eq. (1).

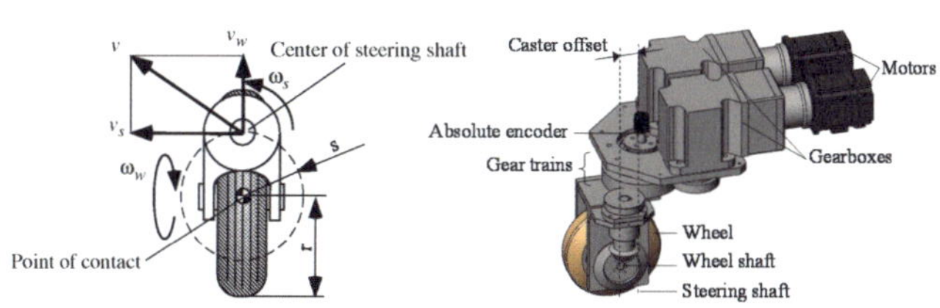

(a) Active-caster (top view) (b) Active-caster drive mechanism

Figure 2. Active-caster. (a) Active-caster (top view). (b) Active-caster drive mechanism.

$$\begin{bmatrix} \omega_w \\ \omega_s \end{bmatrix} = \begin{bmatrix} \frac{\cos\theta}{r} & 0 \\ 0 & \frac{\sin\theta}{s} \end{bmatrix} v \tag{1}$$

where, θ: angle between the velocity vector and wheel,

r: wheel radius,

s: caster offset,

ω_w: rotation of wheel axis,

ω_s: rotation of steering axis, and

v: required velocity vector on the steering shaft.

The active-caster was developed for holonomic and omnidirectional mobile robots by installing two or more numbers of active-caster units on a robotic platform [19]. However, one active-caster unit is used for the electric drive system for propelling the five-wheeled wheelchair whose top view is shown in **Figure 3**. The active-caster generates 2DOF velocity vector whose components are represented by v_s and v_w as shown in **Figure 2(a)**. These components are generated independently by a coordinated control, 2DOF of the wheelchair frame, which can be controlled independently as well. The relationships of the wheelchair motion and the active-caster velocity components are represented in Eq. (2), where x_p is a location of the active-caster steering shaft relative to the midpoint of large wheels of the manual wheelchair as shown in **Figure 3**. From Eqs. (1) and (2), we can derive a control law of the proposed five-wheeled wheelchair as in Eq. (3) to control it in the same manner as a standard electric wheelchair using a joystick.

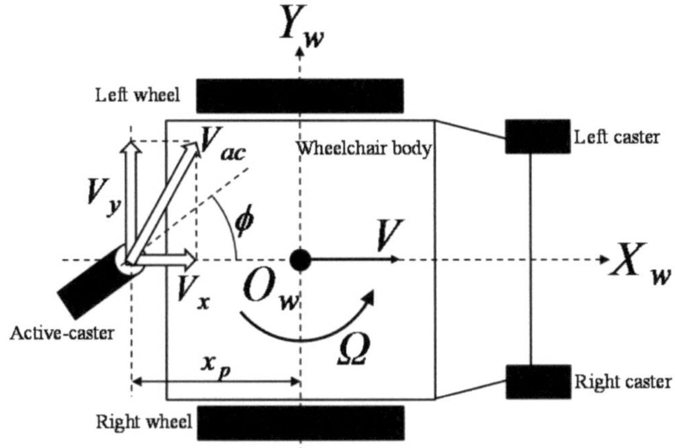

Figure 3. Five-wheeled wheelchair with an add-on active-caster drive wheel.

$$\begin{bmatrix} V_x \\ V_y \end{bmatrix} = \begin{bmatrix} 1 & 0 \\ 0 & x_p \end{bmatrix} \begin{bmatrix} V \\ \Omega \end{bmatrix} \tag{2}$$

$$\begin{bmatrix} \omega_w \\ \omega_s \end{bmatrix} = \begin{bmatrix} \frac{\cos\phi}{r} & \frac{x_p \sin\phi}{r} \\ \frac{-\sin\phi}{s} & \frac{x_p \cos\phi}{s} \end{bmatrix} \begin{bmatrix} V \\ \Omega \end{bmatrix} \tag{3}$$

where V is a translation velocity and Ω is an angular velocity of a wheelchair frame while ϕ is an orientation of the active-caster relative to the wheelchair frame.

2.2. A reconfigurable link mechanism

A five-wheeled wheelchair is composed of a manual wheelchair and the active-caster drive unit. To connect these, we develop a reconfigurable link mechanism to change the location of the active-caster depending on its environments. The major purpose of installing the link mechanism is to surmount a step with no support of caregivers. To vary the location and the height of the active-caster relative to a wheelchair frame, we design a link mechanism whose configuration can be varied by a linear actuator. **Figure 4(a)** shows a schematic side view of the link mechanism with a wheelchair frame and the active-caster. The link mechanism includes three links AB, AC, and BD. The active-caster is installed at point C and touches to the ground. Those links are connected by pin joints at point A, B, and D. The links AC and BD are made of aluminum bars with constant lengths while link AB is composed by cylinder of a linear actuator whose length can be varied by motor actuation. As the length of AB changes, the overall link configuration changes, which results in the change of the location of the active-caster. As the linear actuator extends as shown in **Figure 4(b)**, the active-caster moves backward and the active-caster gets closer to the wheelchair and simultaneously pushes the ground when the linear actuator shrinks as in **Figure 4(c)**.

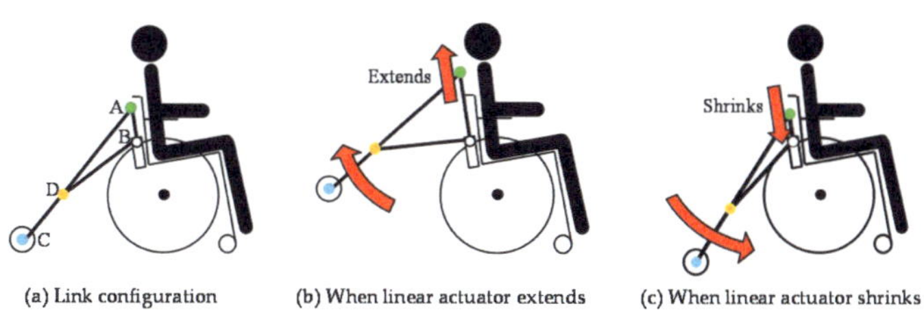

Figure 4. A reconfigurable link mechanism and its motions. (a) Link configuration. (b) When linear actuator extends. (c) When linear actuator shrinks.

3. Step-climbing and -descending strategies for five-wheeled wheelchair

3.1. Step-climbing strategy

The series of motions performed by the wheelchair is shown in **Figure 5**. First, the wheelchair stops in front of the step (**Figure 5(a)**). Next, the wheelchair performs a static wheelie motion and the front casters are hovered from the ground (**Figure 5(b)**). Then, the wheelchair moves forward by maintaining the static wheelie and the front casters reaches on the top of the

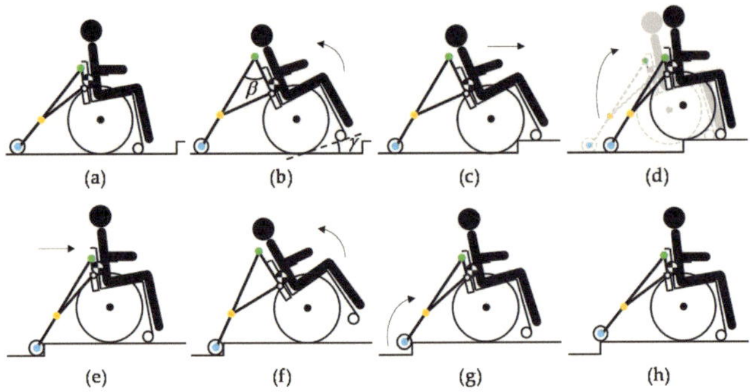

Figure 5. A step-climbing strategy for a five-wheeled wheelchair.

step (**Figure 5(c)**). Note that the center of gravity locates in the area defined by the two large wheels and the drive wheel; therefore, a wheelchair user can move forward with maintaining the posture of the static wheelie without a balancing control. After the front casters climbing, the large wheels are lifted by pushing down the active-caster to the ground (**Figure 5(d)**). By the forward motion, the large wheels pass over the step (**Figure 5(e)**). After the large wheels climbing, the wheelchair performs a static wheelie motion on the step (**Figure 5(f)**). After the completion of the static wheelie, large wheels are locked by the braking mechanism of the wheelchair. By coordinated motions of the drive wheel and the linear actuator (**Figure 5(g)**), the drive wheel climbs up the wall of a step and reaches to the top. Thus, a series of step-climbing sequence is completed (**Figure 5(h)**).

3.2. Step-descending strategy

We then explain the step-descending strategy in which a user approaches a step from the front side of the wheelchair. To avoid a risk of falling from the top of a step to the ground, we apply the static wheelie motion in the step-descending process as well. **Figure 6** shows the series of motions for the step-descending process. First, the wheelchair stops in front of the step (**Figure 6(a)**). Next, the wheelchair performs a static wheelie to hover the front casters and

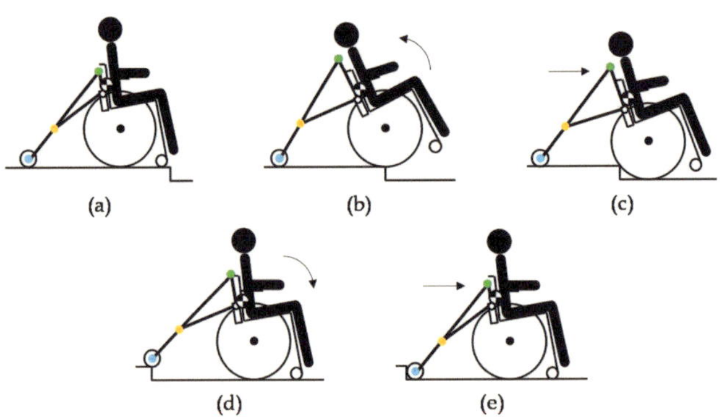

Figure 6. A step-descending strategy for a five-wheeled wheelchair.

moves to the edge of the step (**Figure 6(b)**). After applying gentle brake to the large wheels of a wheelchair for reducing an impact of landing on the ground, the wheelchair starts to descend a step by maintaining the static wheelie situation (**Figure 6(c)**). The breaks are unlocked after the large wheel lands on the ground, the front casters can then be landed on the ground using the linear actuator motion (**Figure 6(d)**). The drive wheel lands on the ground after the forward movement of the wheelchair (**Figure 6(e)**). Thus, the step-descending motion of the wheelchair is completed.

4. Design of a reconfigurable link mechanism

4.1. Geometry model of the reconfigurable link mechanism

In this section, we derive a geometry model of the proposed link mechanism. **Figure 7** shows a schematic side view of the reconfigurable link mechanism with a manual wheelchair. We define a coordinate system of the wheelchair Σ-$X_w Z_w$, where the origin of the coordinate is at the point of contact between the large wheel and the ground as shown in **Figure 7**. The linear actuator on the link mechanism is attached at the back of the wheelchair frame at an angle α. Other links are connected to the linear actuator body, point *B*, and a top of the cylinder, point *A*. Coordinates of points A, C, and D are derived as the relative position from point B.

$$\begin{bmatrix} x_a \\ z_a \end{bmatrix} = \begin{bmatrix} \cos\alpha & -\sin\alpha \\ \sin\alpha & \cos\alpha \end{bmatrix} \begin{bmatrix} 0 \\ u_L \end{bmatrix} + \begin{bmatrix} x_b \\ z_b \end{bmatrix} \tag{4}$$

$$\begin{bmatrix} x_c \\ z_c \end{bmatrix} = \begin{bmatrix} \cos\alpha & -\sin\alpha \\ \sin\alpha & \cos\alpha \end{bmatrix} \begin{bmatrix} -l_0 \sin\beta \\ u_L - l_0 \cos\beta \end{bmatrix} + \begin{bmatrix} x_b \\ z_b \end{bmatrix} \tag{5}$$

$$\begin{bmatrix} x_d \\ z_d \end{bmatrix} = \begin{bmatrix} \cos\alpha & -\sin\alpha \\ \sin\alpha & \cos\alpha \end{bmatrix} \begin{bmatrix} -l_1 \sin\beta \\ u_L - l_1 \cos\beta \end{bmatrix} + \begin{bmatrix} x_b \\ z_b \end{bmatrix} \tag{6}$$

Here, u_L is the length of the link AB, which can be varied by the motion of the linear actuator, $l_{\{0,\,1,\,2\}}$ are the lengths of the links AC, AD, and BD, $x_{\{a,\,b,\,c,\,d\}}$ and $z_{\{a,\,b,\,c,\,d\}}$ are the coordinates of

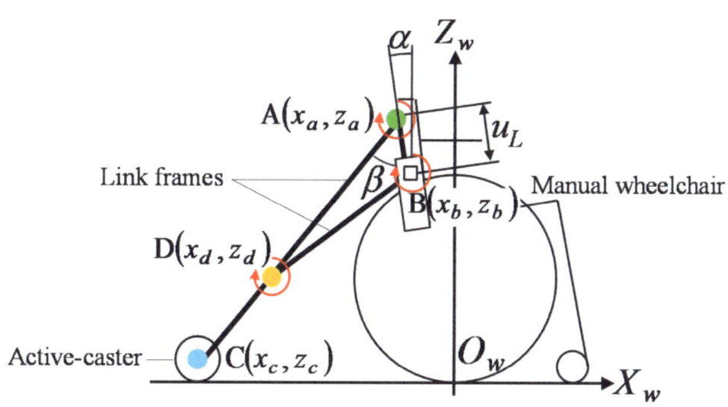

Figure 7. Geometry of a reconfigurable link mechanism.

the points A, B, C, and D along Σ-X_wZ_w axis, respectively. The constant angle α is the inclination of the linear actuator relative to the vertical axis Z_w and β is a variable angle between the link AD and the link AB (∠BAD). We will derive geometric conditions of the link lengths of the mechanism to realize the proposed step-climbing and -descending strategies for the five-wheeled wheelchair.

4.2. Design conditions for climbing up a 100 mm step

Now, we set our target of the surmountable maximum step height to 100 mm, under the assumption that a 100 mm step-climbing capability might allow disable persons to extend their activity in public spots including transfer to trains, passing over steps and gaps in outdoor environments.

At first, we confirm design conditions about the step-climbing of the front casters. In the proposed step-climbing method, the front casters must be hovered 100 mm off the ground for the static wheelie motion. In this motion, the wheelchair frame inclines to the backside. To realize the motion, the link mechanism changes the shape as shown in **Figure 8(a)** in which the active-caster moves backward. The inclination of the wheelchair γ for climbing over a 100 mm high step is represented by the following equation:

$$\gamma = \tan^{-1}\left(\frac{h}{l_f}\right) = 16(\text{deg}) \tag{7}$$

where, h is the height of the step, and l_f is the length between the large wheels and front casters.

Secondly, we confirm the designing condition about the step-climbing of the large wheels. In the proposed method, the large wheels must be hovered 100 mm from the ground. For hovering the motion of the large wheels, the linear actuator changes the height of the drive wheel to locate at 100 mm lower than the ground level as shown in **Figure 8(b)**.

4.3. Design conditions for descending a 100 mm step

Figure 9 shows a schematic view of the wheelchair at a transient moment in which the large wheels descend a step. To prevent wheelchair to fall in the forward direction, a required

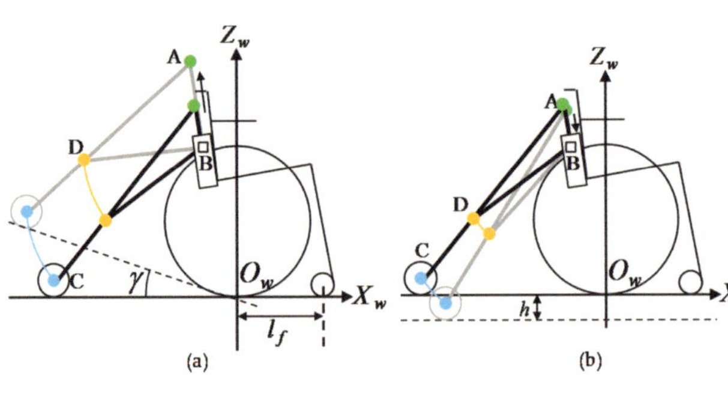

Figure 8. Configurations of a link mechanism. (a) Configuration for static wheelie. (b) Configuration for large wheel hovering.

http://dx.doi.org/10.5772/67558

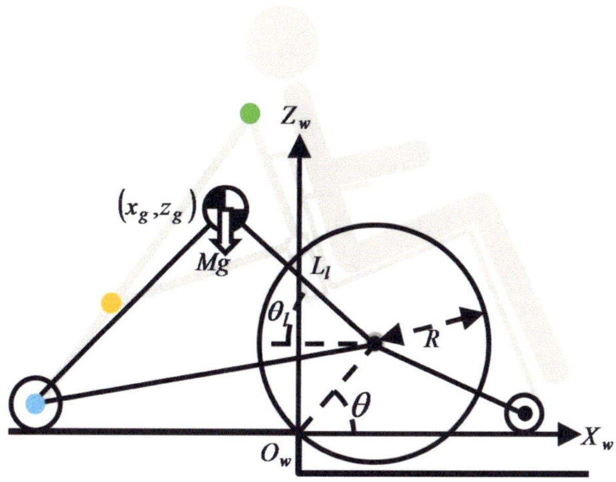

Figure 9. Step-descending model of a five-wheeled wheelchair.

condition for the wheelchair is to maintain the center of gravity not to go beyond the step edge. The condition is represented by a coordinate of the wheelchair along X-axis, x_g as,

$$x_g = R\cos\theta - L_l\cos\theta_l \leq 0 \tag{8}$$

where, R is a radius of the large wheel, L_l is the length between the drive wheel and the center of gravity, and θ_l is the angle between the line L_l and the X_w axis as shown in **Figure 9**.

4.4. Design of the link mechanism

To satisfy the conditions derived above, we determined the parameters of link mechanism for a prototype wheelchair as shown in **Table 1**.

	Symbol	Value	Unit
Length of the links	l_{AC}	0.91	m
	l_{AD}	0.60	m
	l_{BD}	0.51	m
Attachment position	α	8.0	deg
	x_b	−0.13	m
	z_b	0.60	m
Center of gravity (stand condition)	M	124.6	kg
	x_g	−0.054	m
	z_g	0.50	m
Center of gravity (wheelie motion)	x_g	−0.13	m
	z_g	0.62	m

Table 1. Parameters of the prototype wheelchair and the link mechanism.

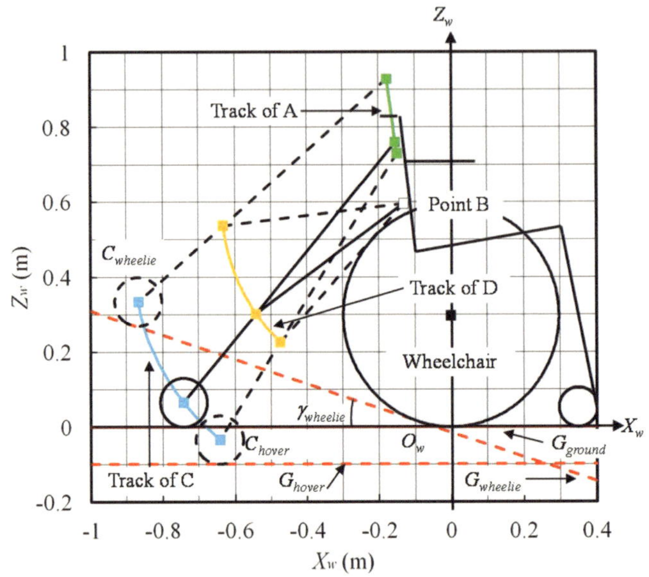

Figure 10. Calculated trajectory of the link mechanism.

By using these parameters, we calculated the motion of the link mechanism. According to the change in the length of link AB, we plotted the trajectories of the points A, C, and D of the proposed mechanism by using Eqs. (4)–(6).

The result is shown in **Figure 10**. As the linear actuator stretches, the drive wheel moves to the back of the wheelchair, by which the inclination γ satisfies the requirement for the static wheelie. Hence as the linear actuator shrinks, the drive wheel gets closer to the wheelchair and pushes the ground for hovering the large wheels to reach a 0.1m high.

5. Semiautomatic control system for step-surmounting motion

In this study, we use laser range finder (LRF) to measure the height of the step and the distance between the wheelchair and the step as shown in **Figure 11(a)**. The measurement result of the step by LRF is shown in **Figure 11(b)**. In **Figure 11(b)**, the positions (i)–(iv) are shown

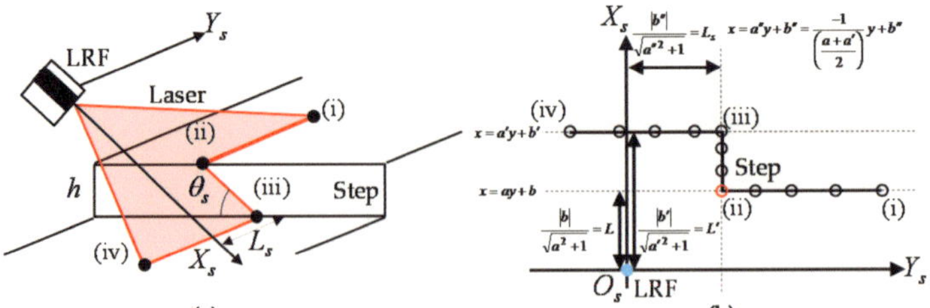

Figure 11. Measurement of the step height and the distance by LRF.

as corresponding to **Figure 11(a)**. The each measurement point by LRF has the coordinates, x and y, on the coordinates system at the center of LRF.

From the LRF data, three lines are identified: the first one represents the top surface of a step, the second one represents the ground, and the last one represents the wall of a step. Those lines are represented as:

$$\begin{aligned} x &= ay + b : \text{step surface} \\ x &= a' y + b' : \text{ground} \\ x &= a'' y + b'' : \text{step wall} \end{aligned} \tag{9}$$

By using constants in these equations, the height of the step, h, and the distance between LRF and the step, L_s, are derived by the following equations:

$$L_s = \frac{|b''|}{\sqrt{a''^2 + 1}} \tag{10}$$

$$h = (L' - L)\sin\theta_s = \left(\frac{|b'|}{\sqrt{a'^2 + 1}} - \frac{|b|}{\sqrt{a^2 + 1}}\right)\sin\theta_s \tag{11}$$

where L and L' are the distances from LRF sensor to a step surface and the ground, respectively. An angle θ_s represents an inclination of a laser scan surface and the ground which can be known as a constant.

By using the measured height of the step h, and the distance to the step L_s, the static wheelie motion and translation motion of the wheelchair can be performed simultaneously to minimize the time for approaching to a step. Moreover, coordinated controls of the linear actuator and the active-caster can be realized for maintaining certain contacts between the large wheels and a step edge for climbing the large wheels.

6. Prototyping

The prototype wheelchair was designed and built whose 3-dimensional Computer Aided Design (3D CAD) models are shown in **Figure 12**. The add-on mechanism, which includes the active-caster and the reconfigurable link mechanism with a linear actuator, is attached on the back of a frame of a manual wheelchair. The stroke of the linear actuator is 200 mm and the maximum power is 144 W. To drive the active-caster, two motors are installed, whose capacities are 200 W each. The diameter of the drive wheel is 130 mm with 45 mm caster offset. Two sensors (LRF) are attached on the side of the wheelchair frame to measure the step whose locations are illustrated in **Figure 13**.

For maneuvering the wheelchair around normal environments, a user can operate the wheelchair by using a joystick on an arm rest as same as in the standard electric wheelchair.

To surmount a step, a user gives a command to a PC (which is not shown in the Figures) to launch the semiautomatic program, which controls the step surmount strategies based on the measured step height and the distance detected by LRFs.

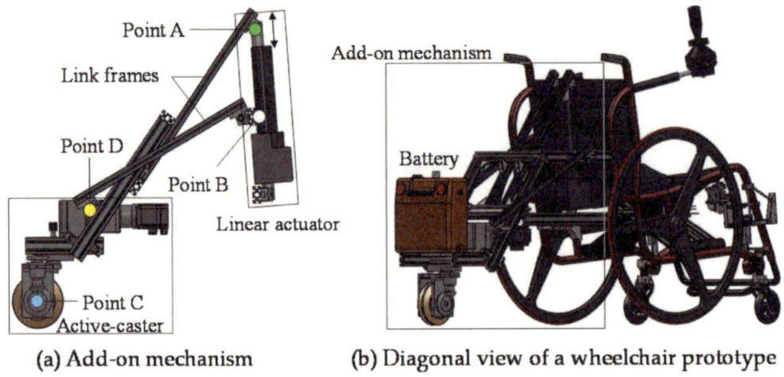

Figure 12. Prototype of a five-wheeled wheelchair with an add-on mechanism. (a) Add-on mechanism. (b) Diagonal view of a wheelchair prototype.

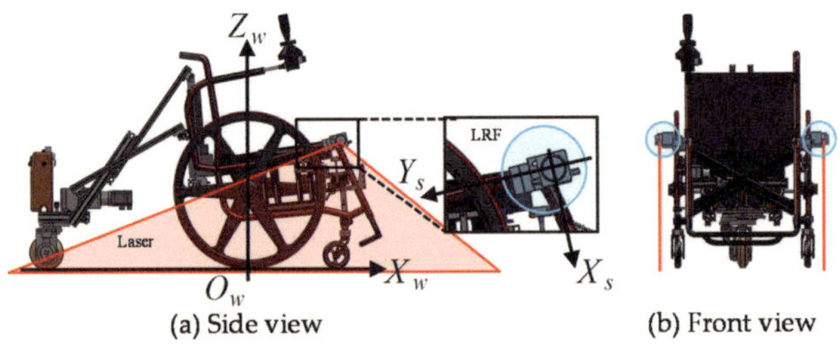

Figure 13. Laser sensors for semiautomatic system. (a) Side view. (b) Front view.

7. Experiment

We propose an add-on electric drive system with a reconfigurable link mechanism for a manual wheelchair. We also propose a semiautomatic system for reducing the user effort to operate a wheelchair to surmount a step. To verify the availability of the prototype wheelchair with the semiautomatic system, we tested the step-climbing and -descending processes of the wheelchair using the prototype. In the experiment, we tested the step-climbing and -descending of a 100 mm step.

7.1. Step climbing

The experimental results are shown in **Figure 14**. In **Figure 14(a)–(h)**, each snapshot of the wheelchair corresponds to **Figure 5(a)–(h)**. The experiments were performed by using the semiautomatic operating system. First, the wheelchair is stopped in front of the step by the user operation using joystick (**Figure 14(a)**). After the user commanded to start the semiautomatic operation, LRFs measured the step height and the distance to the step. Then, the wheelchair approached to the step by performing a static wheelie motion with hovering the front casters not to collide with a step (**Figure 14(b)**). After the large wheels contact the step edge, the wheelchair climbed the step by pushing the active-caster down the ground to lift up the large wheels with maintaining the contacts between the large wheels and the step edge (**Figure 14(c)–(e)**). After the climbing of the large wheels, the static wheelie motion was performed on the top of the step (**Figure 14(f)**). The brake mechanism was then activated after the completion of the static wheelie, for preventing the large wheel to rotate for the drive wheel climbing process. Then the linear actuator changed the link mechanism to locate the drive wheel closer to the wheelchair frame to provide an enough load to the step wall. Together with the linear actuator movements, the drive wheel rotated for climbing the step

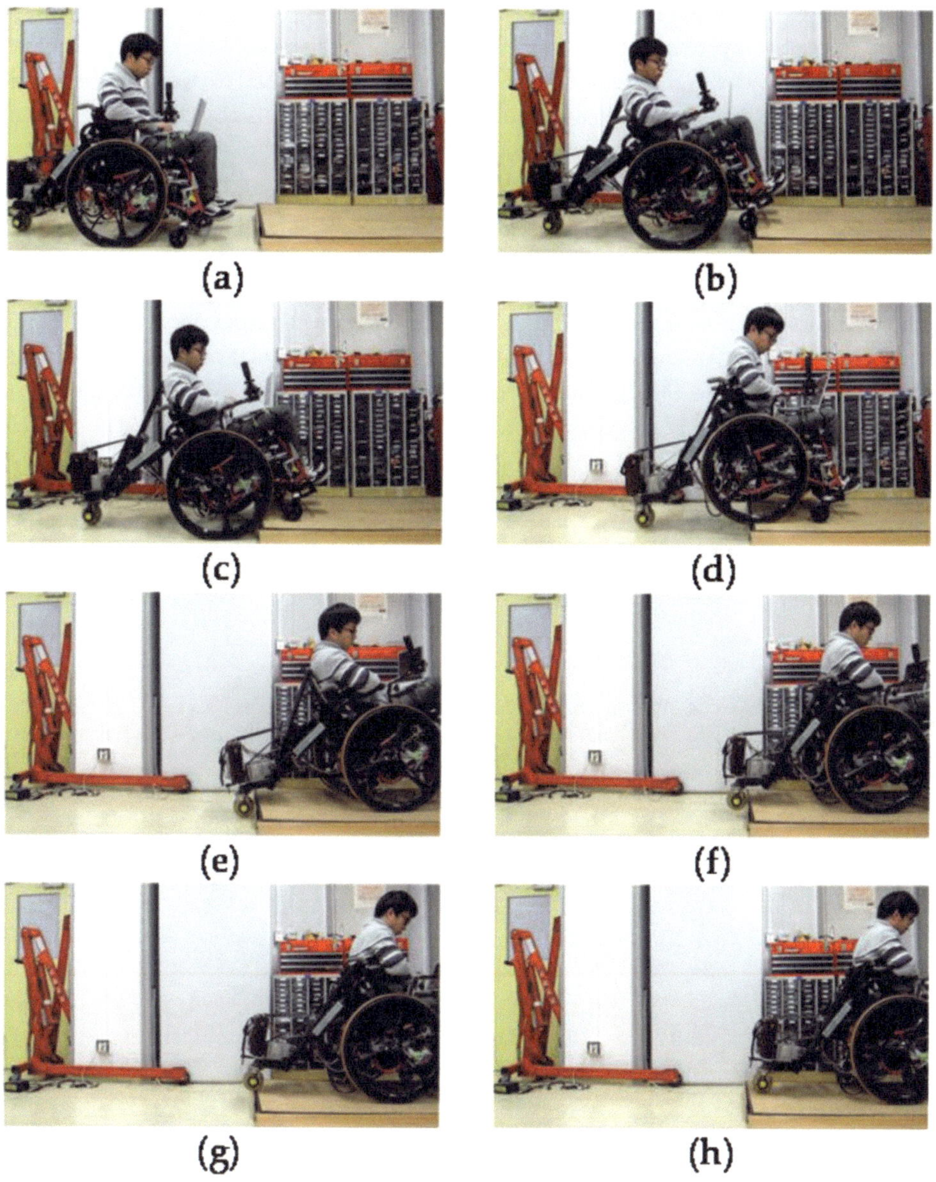

Figure 14. Semiautomatic assist system (step-climbing strategy).

wall (**Figure 14(g)**). Even though the step height was much higher than the diameter of the drive wheel, the drive wheel was successfully able to climb up the step wall since enough load was applied to the step wall by the linear actuator. By unlocking the brake for the large wheels, the step-climbing process was completed (**Figure 14(h)**). From the above all, we confirmed that the wheelchair could climb over a step by the semiautomatic system.

7.2. Step descending

Figure 15 shows the experimental results of the step-descending motion with the semiautomatic operation system. First, the wheelchair stopped in front of the step by the user operation using joystick (**Figure 15(a)**). After the user commanded to start the semiautomatic motion, LRFs measured the step height and the distance between the step edge and the wheelchair. Then, the wheelchair approached to the step by performing a static wheelie motion and stopped to locate the point of contact of the large wheels on the edge of the step (**Figure 15(b)**). After the gentle brakes of the large wheels were activated, the wheelchair descended the step by propelling the drive wheel, which made the large wheels to get down the ground (**Figure 15(c)**). Since the wheelchair performed a static wheelie during the step-descending motion, the user could avoid leaning in the forward direction. After large wheels descending motion, the drive wheel approached to the step edge by contacting the front casters to the ground (**Figure 15(d)**). Finally, the drive wheel reached the ground by decreasing the speed of the drive wheel (**Figure 15(e)**).

In this experiment, the wheelchair was able to descend the step by approaching from the front side of a wheelchair. Thus, by the proposed strategy, we realized the step-descending motion with a reduction of a mental burden.

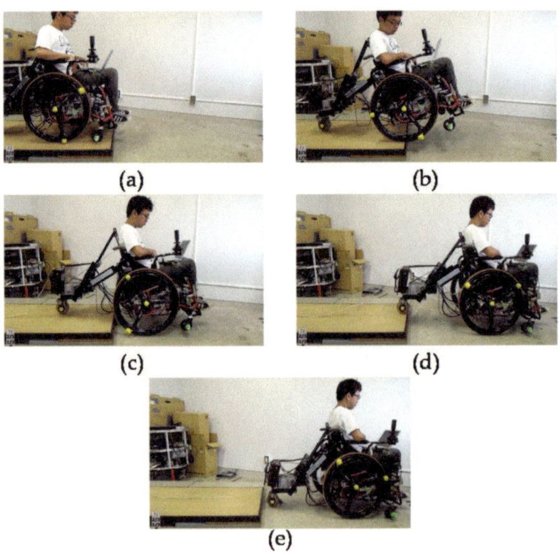

Figure 15. Semiautomatic assist system (step-descending strategy).

8. Conclusion

In this study, we developed a novel add-on mechanism with a reconfigurable link mechanism for a manual wheelchair and realized the step-climbing and -descending strategies. In the climbing motion of the front casters, we proposed the static wheelie motion, in which the large wheels and the drive wheel maintain the static stability with the three-point contact with the ground while the front casters hover from the ground. In the other situation, the large wheels were hovered from the ground by pushing the drive wheel down to the ground using the link mechanism with the linear actuator. We also proposed the climbing motion of the drive wheel by controlling the drive wheel and the linear actuator.

We also proposed the step-descending strategy for approaching a step from the front side of a wheelchair. We derived the design condition of the add-on mechanism for satisfying the static stability of the wheelchair during the proposed descending strategy.

We determined the link length and the position of the add-on mechanism from the design conditions derived from geometric studies. We then confirmed to satisfy the design condition for realizing both of the step-climbing and -descending strategies by link trajectory calculations.

We built the prototype based on the analysis and tested the step-climbing and -descending motions. In the experiments, we confirmed that the proposed wheelchair could climb and descend the step using the proposed strategy and mechanism.

Author details

Masayoshi Wada* and Yu Munakata

*Address all correspondence to: mwada@cc.tuat.ac.jp

Tokyo University of Agriculture and Technology, Koganei-shi, Tokyo, Japan

References

[1] Permobil, Motorized wheelchair, http://www.permobilkk.jp/calling.html. (Supporting Online Materials).

[2] Ulrich Alber GmbH, Scalamobil, https://www.alber.de/. (Supporting Online Materials).

[3] D. Lockton, D. Harrison, "Wheelchair Drive," Project Report for Industrial Design Engineering, Department of Design and Systems Engineering, Brunel University, UK, 2014.

[4] V. Krovi, V. Kumar, "Optimal traction control in a wheelchair with legs and wheels," Proceedings of 4th National Applied Mechanisms and Robotics Conference, 1995.

[5] E. Chaichanasiri, T. Puangumpan, "A prototype of a stair-climbing system for a wheelchair," The Second TSME International Conference on Mechanical Engineering, 2011.

[6] K. Sasaki, Y. Eguchi, K. Suzuki, "Step-climbing wheelchair with lever propelled rotary legs," 2015 IEEE/RSJ International Conference on Intelligent Robots and Systems, pp. 6354-6359, 2015.

[7] N.M. Abdul Ghani, M.O. Tokhi, A.N.K. Nasir, S. Ahmad, "Control of a stair climbing wheelchair," International Journal of Robotics and Automation, 1, 4, pp. 203-213, 2012.

[8] Y. Takahashi, N. Ishikawa, T. Hagiwara, "Soft raising and lowering of front wheels for inverse pendulum control wheelchair robot," Proceedings of the 2003 IEEE/RSJ International Conference on Intelligent Robots and Systems, pp. 3618-3623, 2003.

[9] G. Quaglia, W. Franco, M. Nisi, "Evolution of Wheelchair.q, a stair-climbing wheelchair," The 14th IFToMM World Congress, 2015.

[10] L. Fang, T. Lu, W. He, K. Yuan, "Dynamic and tip-over stability analysis of a planetary wheeled stair-climbing wheelchair," Proceedings of 2012 IEEE International Conference on Mechatronics and Automation, pp. 2541-2546, 2012.

[11] A.S. Shriwaskar, S.K. Choudhary, "Synthesis, modeling and simulation of stair climbing mechanism," International Journal of Engineering Research and Technology, 2, 8, pp. 413-419, 2013.

[12] J. Wang, T. Wang, C. Yao, X-F. Li, C-D. Wu, "Study on interactional characteristics of WT wheelchair robot with stairs during stair-climbing process," Applied Mechanics and Materials, 419, pp. 795-800, 2013.

[13] Segway Inc., iBOT, https://msu.edu/~luckie/segway/iBOT/iBOT.html. (Supporting Online Materials).

[14] Scalevo, http://scalevo.ch/. (Supporting Online Materials).

[15] S. Yokota, D. Chugo, H. Hashimoto, K. Kawabata, "Assistive caster unit of wheelchair for step climbing," International Journal of Mechanical Engineering and Automation, 3(2), pp. 38-45, 2013.

[16] H. Ikeda, et al., "Step climbing and descending for a manual wheelchair with a network care robot," The Second International Conference on Intelligent System and Applications, pp. 95-102, 2013.

[17] Y. Munakata, M. Wada, "A novel step climbing strategy for a wheelchair with active-caster add-on mechanism," IEEE/RSJ International Conference on Intelligent Robots and Systems, pp. 6324-6329, 2015.

[18] M. Wada, S. Mori, "Holonomic and omnidirectional vehicle with conventional tires," Proceedings of the 1996 IEEE International Conference on Robotics and Automation, pp. 3671-3676, 1996.

[19] M. Wada, et al., "Caster drive mechanisms for holonomic and omnidirectional mobile platforms with no over constraint," Proceedings of the International Conference on Robotics and Automation, pp. 1531-1538, 2000.

Chapter 4

Participation and Environmental Factors of Children with Physical Disabilities in Taiwan

Lin-Ju Kang, Ai-Wen Hwang and Chia-Ling Chen

Additional information is available at the end of the chapter

http://dx.doi.org/10.5772/67425

Abstract

Participation is a critical health and education outcome of children and can be optimized by environmental supports. Children with physical disabilities often experience participation restriction and environmental barriers. Research is limited in describing participation in everyday activities of children with physical disabilities and identifying environmental barriers faced by those children in Taiwan. This chapter presents data of 94 children with physical disabilities aged 2–6 years and their families in Taiwan. Children with physical disabilities were primarily children with cerebral palsy (36%) and developmental (motor) delay (34%). Parents completed the Chinese version of Assessment of Preschool Children's Participation (APCP-C) and the Chinese version of the Child and Adolescent Scale of Environment (CASE-C) by structured interview to assess pattern of participation and impact of environment factors to their children's daily life. Participation of children with physical disabilities differed on the basis of level of severity, but not age and sex. Parents reported increased impacts of problems with the quality and availability of family and community resources than problems with assistance/attitude supports and physical design and access. The findings provide a profile of children's pattern of participation and environmental barriers that impact participation in Taiwan.

Keywords: participation, environment, preschool, children, physical disabilities

1. Introduction

The importance of participation and interaction between environment and participation has been emphasized in the International Classification of Functioning, Health, and Disability (ICF) [1]. The ICF provides a conceptual framework for understanding health and well-being of children with physical disabilities [2]. The ICF describes participation as a child's involvement or engagement in life situations and environmental factors such as physical, social, and attitudinal environments that surround an individual. A working definition of participation is "the extent of engagement in the full range of activities that accomplish the larger goal" proposed by Coster and Khetani [3]. In Taiwan, the People with Disabilities Rights Protection Act (2007) states that people with disabilities are viewed as active participants in their lives and participation is the right of all individuals regardless of ability. As supported by the contemporary framework and legislation, participation has been viewed as a critical health and education outcome of children that can be optimized by environmental supports.

Physical disability refers to "any disabilities which limit the physical function of one or more limbs, or movement impairments which limit other facets of daily living." Common congenital and childhood-onset physical disabilities include diagnosis or conditions associated with insults in the central nervous system (CNS), neuromuscular disabilities, or musculoskeletal conditions. These conditions have been used as inclusion criteria for investigation of participation by children with physical disabilities [4–6]. Healthcare services for young children with physical disabilities have traditionally focused on treating impairments (e.g. stretching and strengthening exercises) and training functional skills (e.g. constraint-induced movement training) [7]. Enhancing participation in daily activities through context or environment-based therapies has only recently been considered as a new approach of intervention [8–10].

Emerging research has highlighted participation restrictions and environmental barriers experienced by preschool children with disabilities. Bult et al. indicated that preschool children with physical disabilities participated in fewer play, physical, social, and learning activities and did so less frequently than those without disability [6]. In contrast, Ehrmann et al. reported that preschool children with and without disabilities showed similar patterns of type and frequency of participation in community activities. Children with disabilities, however, participated in fewer family orientated leisure and recreational activities that usually require financial resources compared with peers without disabilities [11]. Khetani et al. reported that children with disabilities demonstrated lower levels of participation frequency, involvement in activities, and parent-perceived environmental supports than those without disabilities, particularly in daycare/preschool settings [12]. Research is limited regarding participation in daily activities and environmental barriers of children less than 6 years of age with physical disabilities in Taiwan.

A multidimensional model of participation of children with physical disabilities would be helpful to understand the complex relationship between participation and environment [13, 14]. Conceptualization of this model was based on contemporary frameworks, empirical evidence, and the ICF model. Based on this model, we proposed that participation encompasses three dimensions: *capability* (i.e. what a child can do in real life), *performance* (i.e. what a child does do), and *subjective experience* (i.e. how a child feels), and the three dimensions of participation are influenced by the determinants of *child, family*, and *environment* (**Figure 1**). This model provides a framework for consideration of dimensions and determinants of participation that are relevant to a child's goals and wishes for full participation. A comprehensive review of current literature was performed in order to understand factors that may influence participation of preschool and school-aged children with physical disabilities. Key determinants of child, family, and environment-related attributes are summarized in **Table 1**.

http://dx.doi.org/10.5772/67425

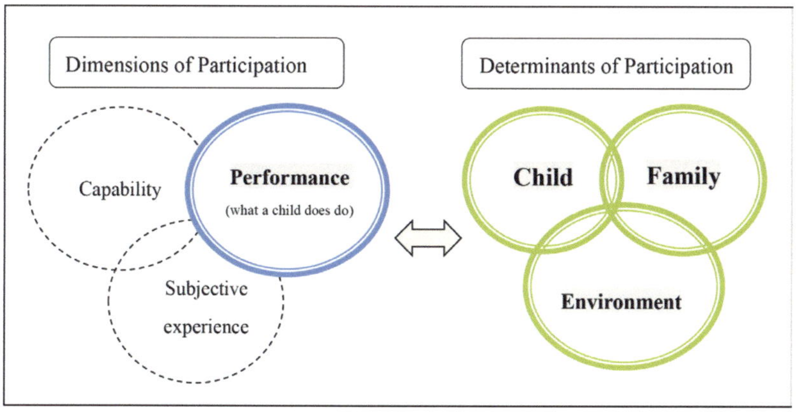

Figure 1. The conceptual framework of this study (adapted from Palisano et al. [13] and Kang et al. [14]).

Dimensions	Attributes for preschool children	Attributes for school-aged children
Child	Age and gender [6, 16, 19]	Age and gender [4, 23–25]
	Severity of impairment [19, 26] Gross motor function [17, 19]	Physical, cognitive, and communicative functioning [4, 23, 25, 27–29].
	Adaptive behaviors [17]	Emotional and behavioral functioning [30]
		Preferences and enjoyment of certain activities, personality, or social orientation [4, 25]
Family	Family income [31]	Family income [24]
	Parental supports and perception of neighborhood safety [31]	Family functioning [4, 25, 32, 33] Family activity preference [25, 32, 33] Parental education [24, 27] Parental stress [24, 27]
Environment	District of residence [34]	Physical environment [35–38]
	Services, support, and attitudes [26]	Social supports [4] Social attitudes [39]
		Services and professional supports [40–42]

Table 1. Summary of determinants of participation identified in literature.

Although current evidence suggests that preschool children with physical disabilities often experience participation restriction and environmental barriers, related research is quite limited in Taiwan. In this chapter, we would like to present findings that are part of a larger longitudinal study investigating determinants of participation of preschool children with physical disabilities. The study is still ongoing and we plan to recruit about 150 preschool children with physical disabilities and 150 age-matched peers with typical development. We will present data from a subsample of children with physical disabilities. First, we will describe patterns of participation in daily activities based on child's age, sex, and level of severity. Second, we will present environmental barriers that impact participation identified by parents of those children with physical disabilities.

2. Patterns of participation of children

2.1. Research methods and materials for measuring participation

Data on 94 children with physical disabilities aged 2–6 years and their families were collected from various regions in Taiwan. This sample was also used for establishing the reliability and validity of the key participation measure (described below) used in this study [15]. Children had a mean age of 4.2 years (SD = 1.4), 67% were boys, and 73% attended preschool or developmental centers. Children with physical disabilities were primarily children with cerebral palsy (36%) and developmental (motor) delay (34%), followed by chromosomal disorder (15%), acquired brain injury (13%), and congenital anomalies (2%). Medical diagnoses and level of severity (mild, moderate, severe, profound) were determined by children's physicians in the certified hospitals and reported by parents. Parent respondents were primarily mothers (84%), followed by fathers (14%) and grandparents (2%). Parents completed the Chinese version of Assessment of Preschool Children's Participation (APCP-C) by structured interview.

The APCP-C is a measure of participation in play, skill development, active physical recreation, and social activities of preschool children with and without disabilities. The English version of the APCP was developed and validated for children with cerebral palsy in Canada [16]. The APCP was also used in several studies involving young children with CP in the United States, Canada, and Taiwan [8, 17–19] and young children with physical disabilities in the Netherlands [6]. The APCP-C includes 45 activities, and for each activity, a parent indicates whether the child has performed the activity over the past 4 months ("yes" or "no"). If yes, the parent then reports how often the child performed the activity on a seven-point Likert scale (1 = "once over the past 4 months" to 7 = "once daily or more"). Diversity and intensity scores were calculated for total scores (all items) and for each activity type. A diversity score was the sum of the total number of activities performed, and an intensity score is the sum of frequencies for all items divided by the number of possible items for all items as well as items in each activity type. Evidence of internal consistency (Cronbach's α = 0.54–0.86) and test-retest reliability (ICCs = 0.56–0.79), cultural validity, and convergent validity has been reported for the APCP-C diversity and intensity scores [15].

2.2. Describing participation by age, sex, and severity

Participation of children with physical disabilities differed on the basis of level of severity in impairments, but not age and sex (**Tables 2–4**). Children younger and older than 4 years, and boys and girls did not differ in participation diversity and intensity across all types of activities ($p > 0.05$) (**Tables 2** and **3**). Significant differences in participation diversity and intensity in the total and three activity types other than play activities were found between children with different levels of severity (**Table 4**). Children with a higher level of severity had a lower level of participation diversity and intensity. *Post hoc* comparisons revealed a similar pattern across all types of activities. Children who were classified as mild level had significantly higher total participation diversity and intensity than children at moderate, severe, or profound levels. Children who were classified as mild level had significantly higher participation

Age group	<4 years $n = 43$	≥4 years $n = 51$	t^a	p^a (two-tailed)
Total				
Diversity	23.1 (7.3)	26.5 (7.8)	−2.20	0.03
Intensity	2.6 (0.8)	2.8 (0.9)	−1.17	0.24
Play				
Diversity	5.7 (1.6)	6.1 (1.7)	−1.11	0.27
Intensity	3.7 (1.1)	3.7 (1.1)	0.23	0.81
Skill development				
Diversity	7.4 (3.2)	9.3 (3.5)	−2.61	0.01
Intensity	2.5 (1.2)	3.1 (1.2)	−2.32	0.02
Active physical recreation				
Diversity	5.1 (1.8)	5.4 (2.0)	−0.83	0.41
Intensity	2.7 (0.9)	2.6 (1.0)	0.38	0.70
Social				
Diversity	4.7 (1.9)	5.67 (2.1)	−2.04	0.04
Intensity	1.8 (0.8)	2.08 (0.8)	−1.57	0.11

[a]Independent t tests (significance level was set as $p < 0.01$ due to number of comparisons performed).
Scores are presented as means (SD).

Table 2. Participation diversity and intensity based on child's age.

intensity than children at profound level in skill development, active physical recreation, and social activities. Participation diversity between children classified as mild severity and children classified as severe level differed only in skill development activities.

Children with physical disabilities aged under or above 4 years of age did not differ in the levels of participation in everyday activities. Our findings were supported by the study of Chiarello et al. in which they found no differences in amount of participation among 3-, 4-, and 5-year-old children with CP in the United States. In contrast, both the Canadian and Dutch studies found that the APCP distinguished participation of children with CP and physical disabilities under and above 4 years of age [6, 16]. Collectively, the inconsistent results from these studies suggest that age effects on patterns of participation at preschool years are culturally sensitive and in need for further study.

Boys and girls did not differ in the levels of participation across all types of activities. The results indicate that in a Taiwanese culture, boys and girls have not yet developed divergent interests and activities at this early developmental age. Boys and girls were probably given similar activity opportunities by their adult caregivers at home, preschool, or community. The social impact of sex may not play an important role in affecting participation of preschool children. Previous studies also showed inconsistent results regarding the sex effect on

Sex	Boys $n = 63$	Girls $n = 31$	t^a	p^a (two-tailed)
Total				
Diversity	25.5 (6.8)	23.8 (9.2)	1.04	0.30
Intensity	2.8 (0.8)	2.6 (1.0)	1.07	0.28
Play				
Diversity	6.1 (1.6)	5.5 (1.8)	1.41	0.16
Intensity	3.8 (0.9)	3.6 (1.3)	0.68	0.49
Skill development				
Diversity	8.5 (3.1)	8.2 (4.2)	0.38	0.70
Intensity	2.8 (1.2)	2.8 (1.3)	0.25	0.80
Active physical recreation				
Diversity	5.5 (1.7)	4.7 (2.3)	1.84	0.06
Intensity	2.8 (0.9)	2.4 (1.0)	1.71	0.09
Social				
Diversity	5.3 (2.0)	5.1 (2.2)	0.34	0.73
Intensity	1.9 (0.8)	1.9 (0.8)	0.29	0.77

[a]Independent t tests (significance level was set as $p < 0.01$ due to number of comparisons performed).
Scores are presented as means (SD).

Table 3. Participation diversity and intensity based on child's sex.

Severity	Mild $n = 23$	Moderate $n = 14$	Severe $n = 21$	Profound $n = 14$	F^a	p^a (two-tailed)	*Post hoc* comparisons[b]
Total							
Diversity	30.1 (6.2)	24.0 (9.0)	24.5 (7.0)	20.7 (5.4)	5.97	0.001	Mild > severe ($p = 0.047$) Mild > profound ($p = 0.001$)
Intensity	3.4 (0.7)	2.5 (0.9)	2.6 (0.8)	2.2 (0.6)	8.09	<0.001	Mild > moderate ($p = 0.012$), mild > severe ($p = 0.007$), mild > profound ($p < 0.001$)
Play							
Diversity	6.7 (1.5)	5.5 (1.9)	5.9 (1.6)	5.0 (1.8)	3.34	0.02	-
Intensity	4.1 (1.0)	3.3 (1.2)	3.7 (1.0)	3.2 (1.0)	2.80	0.04	-
Skill development							
Diversity	10.8 (3.2)	8.5 (3.6)	7.5 (3.2)	6.7 (2.3)	6.31	0.001	Mild > severe ($p = 0.005$) Mild > profound ($p = 0.001$)

Severity	Mild $n = 23$	Moderate $n = 14$	Severe $n = 21$	Profound $n = 14$	F^a	p^a (two-tailed)	*Post hoc* comparisons[b]
Intensity	3.7 (0.9)	2.8(1.3)	2.3 (1.2)	2.1 (0.9)	8.47	<0.001	Mild > severe ($p = 0.001$) Mild > profound ($p < 0.001$)
Active physical recreation							
Diversity	6.3 (1.3)	5.0 (2.8)	5.3 (1.7)	4.5 (1.9)	2.93	0.04	-
Intensity	3.3 (0.7)	2.4 (1.1)	2.5 (0.9)	2.2 (0.8)	5.09	0.003	Mild > profound ($p = 0.004$)
Social							
Diversity	6.2 (2.0)	4.9 (2.4)	5.7 (1.9)	4.4 (1.4)	2.82	0.04	-
Intensity	2.3 (0.8)	1.7 (0.9)	2.0 (0.7)	1.4 (0.5)	4.42	0.007	Mild > profound ($p = 0.005$)

[a]ANOVA tests (significance level was set as $p < 0.01$ due to number of comparisons performed).
[b]Tukey HSD tests (significance level was set as $p < 0.05$).
Scores are presented as means (SD).

Table 4. Participation diversity and intensity based on child's level of severity.

preschool participation. The US study of the APCP reported no sex differences [17], whereas the Canadian and Dutch studies found that girls had higher intensity of participation in play [16] and skill development activities. It is worth of further studying of whether sex plays an important role in differentiating participation in this early period of life.

Children with mild severity as classified by their physicians were found to have higher participation intensity in skill development, active physical recreation, and social activities than children with profound severity. Skill development, active physical recreation, and social activities involve physical, cognitive, and social demands of the child and may require preparation and assistance from adults. Children with profound level of severity may experience particular difficulties even with activity adaptation or modification. Participation in play activities did not differ among children across the levels of severity. The result made practical sense, given that play activities are more easily adapted than other type of activities and use materials that are easily accessible at home (e.g. toys, TV, or other household items).

3. Environmental barriers

3.1. Research methods and materials for measuring environment

As described in the above section, the same set of data of 94 children with physical disabilities is reported. Parents completed the Chinese version of the Child and Adolescent Scale of Environment (CASE-C) by structured interview to assess impact of environment factors to their children's daily life.

The CASE-C is a measure of the impact of environmental features to the child's home, school, and community, which contains three subscales: family/community resources, assistance/attitude supports, and physical design access problems [20]. The English version of the CASE was developed and validated for children with acquired brain injury and various chronic conditions in the United States [21, 22]. The first 18 items of the CASE-C were close-ended questions, and each item is rated on a three-point scale: no problem (1), little problem (2), and big problem (3). The CASE-C scores were calculated for total scores (all items) and for each subscale then adjusted to a 0 to 100-point scale. Higher scores indicate a greater impact of environmental problems. The 19th item is an open-ended question asking parents to specify special environmental conditions that are considered as important barriers. Parents' qualitative answers of the 19th item were classified by content analysis based on the five ICF domains of environmental factors: products and technology (e1), natural and human-made changes (e2), support and relationship (e3), attitude (e4), and services, systems, and policies (e5) [1].

3.2. Impacts of environmental barriers

Parents of children with physical disabilities perceived a low level of environmental problems to their children's participation in home, school, and community, as the CASE-C scores ranged from 40.8 to 44.3 of 100 (**Table 5**). Parents reported increased impacts of problems with the quality and availability of family and community resources, including lack of community programs and services, inadequate or lack of information about child's diagnosis or intervention, and problems with services and policies provided by government agencies.

CASE-C scale/subscale	Scores
Total scale	42.6 (8.9)
Family/community resources	44.3 (11.6)
Assistance/attitude supports	41.8 (12.2)
Physical design access problems	40.8 (8.9)
Scores are presented as means (SD).	

Table 5. Scores for impacts of environmental barriers.

Twenty parents specified a total of 36 environmental conditions that were important barriers (**Table 6**), over half of the conditions pertained to the domain of services, systems, and policies (e5) (56%). The results again support the relatively higher perceived barriers of family and community resources.

In particular, we had the impression from interviews with families that parents often felt nervous and helpless during their children's transition to elementary school due to rapid decrease of therapy services. In Taiwan, early intervention services are provided for children less than 6 years of age. Many rehabilitation services are child-focused, and parents are not actively involved in the intervention process. When children reach school age and the early intervention services are no longer available, their parents are anxious about not getting help

ICF domains	Count (%)	Summary of conditions specified by parents
e1 Products and technology e2 Natural environment and human-made changes to environment	8 (22%)	• Insufficiency of accessible facilities in the public areas, for instance, restrooms for parents with young children are unavailable in the park, toilets are too high, and buses are inaccessible for a child with physical disability. • Some mother resigned from work to look after their children, resulting in huge financial burden. They even could not afford products for daily living (such as diapers) and medical expenses.
e3 Support and relationships	3 (8%)	• Parents expressed the difficulties to understand their therapists because of the professional terminologies. • Peers sometimes refuse to play with the child because they do not understand the child's condition and limitations.
e4 Attitudes	5 (14%)	• Parents felt uncomfortable by the ways people looking at them while going out with their child. • Unfriendly service providers such as clinicians in the hospitals or taxi drivers.
e5 Services, systems and policies	20 (56%)	• Information was either unavailable or lack of integration. Parents usually did not know how and where to find information they need. • It was difficult for parents to apply for subsidy, because of restrictions set by the government. • Some families need to find a place where children can play safely very far away from their home. • Lack of therapy services in the nearby area. • Parents were not allowed to enter therapy rooms during their child's therapy sessions; parents did not know their child's performance. • During transition to elementary school, the therapy services and resources were no longer available for some families. • School professionals did not have sufficient competence in assisting children in engaging in classes. • Lack of opportunity for children with disabilities to receive inclusion education in mainstream schools.

Table 6. Parent-reported environmental conditions (total count = 36) classified by the ICF domains of environmental factors.

from school professionals. We suggest that parents are encouraged to be actively involved in early intervention services and be empowered so they have the competence to advocate for their children.

4. Conclusions

The findings provide a snapshot of children's participation and environment in Taiwan. Participation of children with physical disabilities did not vary by child's age and sex.

Children with severe or profound levels of severity tend to have lower diversity and intensity of participation than children with mild level of severity. Inadequate or lack of family and community resources had greater impacts than problems with support, assistance, and attitudes and physical design and access. Environmental barriers identified in our study will provide suggestions to the clinical practice and government policies to improve environmental accessibility, information availability, and integration and to support children's integration in society.

Acknowledgements

The authors would like to thank the families involved in this study. Drs. Robert J. Palisano, Lisa A. Chiarello, and Gillian A. King are thanked for their constructive feedback for conceptualization and data interpretation of this study. Ms. Yi-Chen Li and Ming-Chieh Hsieh are thanked for their assistance in data collection and analysis. The Ministry of Science and Technology [MOST 103-2314-B-182-001 & 104-2314-B-182-022-MY3] and Chang Gung Memorial Hospital [CMRPD1D0111] in Taiwan are appreciated for financially supporting this research.

Author details

Lin-Ju Kang*, Ai-Wen Hwang and Chia-Ling Chen

*Address all correspondence to: linjukang@mail.cgu.edu.tw

1 Graduate Institute of Early Intervention, Chang Gung University, Linkou, Taoyuan, Taiwan

2 Department of Physical Medicine Rehabilitation, Chang Gung Memorial Hospital, Linkou, Taoyuan, Taiwan

References

[1] World Health Organization. The International Classification of Functioning, Disability and Health. 2001. ICF: Geneva: World Health Organization.

[2] Palisano RJ, Snider LM, Orlin MN. Recent advances in physical and occupational therapy for children with cerebral palsy. Seminars in Pediatric Neurology. 2004;**11**:66-76.

[3] Coster W, Khetani MA. Measuring participation of children with disabilities: Issues and challenges. Disability and Rehabilitation. 2008;**30**:639-48. DOI: 10.1080/09638280701400375.

[4] King G, Law M, Hanna S, King S, Hurley P, Rosenbaum P. Predictors of the leisure and recreation participation of children with physical disabilities: A structural equation modeling analysis. Children's Health Care. 2006;**35**(3):209-34. DOI: 10.1207/s15326888chc3503_2.

[5] Bult MK, Verschuren O, Gorter JW, Jongmans MJ, Piskur B, Ketelaar M. Cross-cultural validation and psychometric evaluation of the Dutch language version of the Children's Assessment of Participation and Enjoyment (CAPE) in children with and without physical disabilities. Clinical Rehabilitation. 2010;**24**(9):843-53. DOI: 10.1177/0269215510367545.

[6] Bult MK, Verschuren O, Kertoy MK, Lindeman E, Jongmans MJ, Ketelaar M. Psychometric evaluation of the Dutch version of the Assessment of Preschool Children's Participation (APCP): Construct validity and test-retest reliability. Physical and Occupational Therapy in Pediatrics. 2013;**33**(4):372-83. DOI: 10.3109/01942638.2013.764958.

[7] Campbell SK, Vander Linden DW, Palisano RJ. Physical therapy for children. 2006. 3rd ed. Philadelphia, PA: Saunders Elsevier Inc..

[8] Law M, Darrah J, Pollock N, Wilson B, Russell DJ, Walter SD, et al. Focus on function: A cluster, randomized controlled trial comparing child-versus context-focused intervention for young children with cerebral palsy. Developmental Medicine and Child Neurology. 2011;**53**(7):621-9. DOI: 10.1111/j.1469-8749.2011.03962.x.

[9] Darrah J, Law MC, Pollock N, Wilson B, Russell DJ, Walter SD, et al. Context therapy: A new intervention approach for children with cerebral palsy. Developmental Medicine and Child Neurology. 2011;**53**(7):615-20. DOI: 10.1111/j.1469-8749.2011.03959.x.

[10] Anaby DR, Law MC, Majnemer A, Feldman D. Opening doors to participation of youth with physical disabilities: An intervention study. Canadian Journal of Occupational Therapy. 2016;**83**(2):83-90. DOI: 10.1177/0008417415608653.

[11] Ehrmann LC, Aeschleman SR, Svanum S. Parental reports of community activity patterns: A comparison between young children with disabilities and their nondisabled peers. Research in Developmental Disabilities. 1995;**16**(4):331-43. DOI: 10.1016/0891-4222(95)00017-H.

[12] Khetani MA, Graham JE, Davies PL, Law MC, Simeonsson RJ. Psychometric properties of the young children's participation and environment measure. Archives of Physical Medicine and Rehabilitation. 2015;**96**(2):307-16. DOI: 10.1016/j.apmr.2014.09.031.

[13] Palisano RJ, Chiarello LA, King G, Novak I, Stoner T, Fiss A. Participation-based therapy for children with physical disabilities. Disability & Rehabilitation. 2012;**34**(12):1041-52. DOI: 10.3109/09638288.2011.628740.

[14] Kang LJ, Palisano RJ, King GA, Chiarello LA. A multidimensional model of optimal participation of children with physical disabilities. Disability and Rehabilitation. 2014;**36**(20):1735-41. DOI: DOI: 10.3109/09638288.2013.863392.

[15] Kang LJ, Hwang AW, Palisano RJ, King GA, Chiarello LA, Chen CL. Validation of the Chinese version of the Assessment of Preschool Children's Participation for children with physical disabilities. Developmental Neurorehabilitation. 2016:1-8. DOI: 10.3109/17518423.2016.1158746.

[16] Law M, King G, Petrenchiki T, Ketrtoy M, Anaby D. The Assessment of Preschool Children's Participation: Internal consistency and construct validity. Physical & Occupational Therapy in Pediatrics. 2012;**32**(3):272-87. DOI: 10.3109/01942638.2012.662584.

[17] Chiarello LA, Palisano RJ, Orlin MN, Chang HJ, Begnoche D, An M. Understanding participation of preschool-age children with cerebral palsy. Journal of Early Intervention. 2012;**34**(1):3-19.

[18] Bjornson KF, Zhou C, Stevenson R, Christakis DA. Capacity to participation in cerebral palsy: Evidence of an indirect path via performance. Archives of Physical Medicine and Rehabilitation. 2013;**94**(12):2365-72. DOI: 10.1016/j.apmr.2013.06.020.

[19] Wu KP, Chuang YF, Chen CL, Liu IS, Liu HT, Chen HC. Predictors of participation change in various areas for preschool children with cerebral palsy: A longitudinal study. Research in Developmental Disabilities. 2015;**37**:102-11. DOI: 10.1016/j.ridd.2014.11.005.

[20] Kang LJ, Yen CF, Bedell G, Simeonsson RJ, Liou TH, Chi WC, et al. The Chinese version of the Child and Adolescent Scale of Environment (CASE-C): Validity and reliability for children with disabilities in Taiwan. Research in Developmental Disabilities. 2015;**38**:64-74. DOI: 10.1016/j.ridd.2014.12.019.

[21] Bedell G Developing a follow-up survey focused on participation of children and youth with acquired brain injuries after discharge from inpatient rehabilitation. NeuroRehabilitation. 2004;**19**(3):191-205.

[22] Bedell G, McDougall J. The Child and Adolescent Scale of Environment (CASE): Further validation with youth who have chronic conditions. Developmental Neurorehabilitation. 2015;**18**(6):375-82. DOI: 10.3109/17518423.2013.855273.

[23] Bult MK, Verschuren O, Jongmans MJ, Lindeman E, Ketelaar M. What influences participation in leisure activities of children and youth with physical disabilities? A systematic review. Research in Developmental Disabilities. 2011;**32**:1521-1529. DOI: 10.1016/j.ridd.2011.01.045.

[24] Law M, King G, King S, Kertoy M, Hurley P, Rosenbaum P, et al. Patterns of participation in recreational and leisure activities among children with complex physical disabilities. Developmental Medicine and Child Neurology. 2006;**48**(5):337-42. DOI: 10.1017/S0012162206000740.

[25] Palisano RJ, Orlin M, Chiarello LA, Oeffinger D, Polansky M, Maggs J, et al. Determinants of intensity of participation in leisure and recreational activities by youth with cerebral palsy. Archives of Physical Medicine and Rehabilitation. 2011;**92**(9):1468-76. DOI: 10.1016/j.apmr.2011.04.007.

[26] Forsyth R, Colver A, Alvanides S, Woolley M, Lowe M. Participation of young severely disabled children is influenced by their intrinsic impairments and environment. Developmental Medicine and Child Neurology. 2007;49:345-9. DOI: 10.1111/ j.1469-87492007.00345.x.

[27] Majnemer A, Shevell M, Law M, Birnbaum R, Chilingaryan G, Rosenbaum P, et al. Participation and enjoyment of leisure activities in school-aged children with cerebral palsy. Developmental Medicne and Child Neurology. 2008;**50**:751-8. DOI:10.1111/j.1469-8749.2008.03068.x.

[28] Donkervoort M, Roebroeck M, Wiegerink D, van der Heijden-Maessen H, Stam H, The Transition Research Group South West N. Determinants of functioning of adolescents and young adults with cerebral palsy. Disability & Rehabilitation. 2007;**29**(6):453-63. DOI: 10.1080/09638280600836018.

[29] Brown M, Gordon WA. Impact of impairment on activity patterns of children. Archives of Physical Medicine and Rehabilitation. 1987;**68**(12):828-32.

[30] King G, McDougall J, DeWit D, Hong S, Miller L, Offord DR, et al. Pathways to children's academic performance and prosocial behaviour: Roles of physical health status, environmental, family, and child factors. International Journal of Disability, Development and Education. 2005;**52**(4):313-44. DOI: 10.1080/10349120500348680.

[31] Carlson E, Bitterman A, Daley T. Access to Educational and Community Activities for Young Children with Disabilities: Selected Findings from the Pre-Elementary Education Longitudinal Study (PEELS). 2010. Retrieved from http://ies.ed.gov/ncser/pubs/20113000/pdf.

[32] Bennett KS, Hay DA. The role of family in the development of social skills in children with physical disabilities. International Journal of Disability, Development and Education. 2007;**54**(4):381-97. DOI: 10.1080/10349120701654555.

[33] McGee R, Williams S, Howden-Chapman P, Martin J, Kawachi I. Participation in clubs and groups from childhood to adolescence and its effects on attachment and self-esteem. Journal of Adolescence. 2006;**29**(1):1-17. DOI: 10.1016/j.adolescence.2005.01.003.

[34] Hammal D, Jarvis SN, Colver AF. Participation of children with cerebral palsy is influenced by where they live. Developmental Medicine and Child Neurology. 2004;**46** (5):292-8.

[35] Barf HA, Post MW, Verhoef M, Jennekens-Schinkel A, Gooskens RH, Prevo AJ. Restrictions in social participation of young adults with spina bifida. Disability and Rehabilitation. 2009;**31**(11):921-7. DOI: 10.1080/09638280802358282.

[36] Mihaylov SI, Jarvis SN, Colver AF, Beresford B. Identification and description of environmental factors that influence participation of children with cerebral palsy. Developmental Medicne and Child Neurology. 2004;**46**(5):299-304. DOI: 10.1017/S0012162204000490.

[37] Shikako-Thomas K, Majnemer A, Law M, Lach L. Determinants of participation in leisure activities in children and youth with cerebral palsy: Systematic review. Physical & Occupational Therapy in Pediatrics. 2008;**28**(2):155-69.

[38] Chan HS, Lau PH, Fang KH, Poon D, Lam CC. Neuroimpairment, activity limitation, and participation restriction among children with cerebral palsy in Hong Kong. Hong Kong Medical Journal. 2005;**11**:342-50.

[39] Anaby D, Hand C, Bradley L, Direzze B, Forhan M, Digiacomo A, et al. The effect of the environment on participation of children and youth with disabilities: A scoping review. Disability and Rehabilitation. 2013;**35**(19):1589-98. DOI: 10.3109/09638288.2012.748840.

[40] King G.. The meaning of life experiences: Application of a meta-model to rehabilitation sciences and services. American Journal of Orthopsychiatry. 2004;**74**(1):72-88.

[41] Rosenbaum P, King S, Law M, King G, Evans J. Family-centred service: A conceptual framework and research review. Physical Occupational Therapy in Pediatrics. 1998;**18**:1-20. DOI: 10.1080/J006v18n01_01.

[42] Law M, Haight M, Milroy B, Willms D, Stewart D, Rosenbaum P. Environmental factors affecting the occupations of children with physical disabilities. Journal of Occupational Science. 1999;**6**(3):102-10. DOI: 10.1080/14427591.1999.9686455.

Chapter 5

Hand Rehabilitation after Chronic Brain Damage: Effectiveness, Usability and Acceptance of Technological Devices: A Pilot Study

Marta Rodríguez-Hernández,

Carmen Fernández-Panadero,

Olga López-Martín and Begoña Polonio-López

Additional information is available at the end of the chapter

http://dx.doi.org/10.5772/67532

Abstract

Purpose: The aim is to present an overview of existing tools for hand rehabilitation after brain injury and a pilot study to test HandTutor® in patients with chronic brain damage (CBD).

Method: Eighteen patients with CBD have been selected to test perception on effectiveness, usability and acceptance of the device. This group is a sample of people belonging to a wider study consisting in a randomized clinical trial (RCT) that compares: (1) experimental group that received a treatment that combines the use of HandTutor® with conventional occupational therapy (COT) and (2) control group that receives only COT.

Results: Although no statistical significance has been analysed, patients report acceptance and satisfaction with the treatment, decrease of muscle tone, increase of mobility and better performance in activities of daily life. Subjective perceptions have been contrasted with objective measures of the range of motion before and after the session. Although no side effects have been observed after intervention, there has been some usability problems during setup related with putting on gloves in patients with spasticity.

Conclusions: This chapter is a step further of evaluating the acceptance of technological devices in chronic patients with CBD, but more research is needed to validate this preliminary results.

Keywords: HandTutor®, rehabilitation, chronic brain damage, stroke, traumatic brain injury

1. Introduction

According to the World Health Organization [1], cerebrovascular accidents (stroke) are the second leading cause of death and the third leading cause of disability. The last update of the global Burden of Ischemic and Haemorrhagic Stroke [2] indicates that although age-standardized rates of stroke mortality have decreased worldwide in the past two decades, the absolute numbers of people who have a stroke every year are increasing. In 2013, there were 10.3 million of new strokes, 6.5 million deaths from stroke, almost 25.7 million stroke survivors and 113 million of people with disability-adjusted life years (DALYs) due to stroke.

One of the most frequent problems after stroke is upper limb (UL) impairments such as muscle weakness, contractures, changes in muscle tone, and other problems related to coordination of arms, hands or fingers [3, 4]. These impairments induce disabilities in common movements such as reaching, picking up or holding objects and difficult activities of daily living (ADLs) such as washing, eating or dressing, their participation in society, and their professional activities [5]. Most of people experiencing this upper limb impairment will still have problems chronically several years after the stroke. Impairment in the upper limbs is one of the most prevalent consequences of stroke. For this reason making rehabilitation is an essential step towards clinical recovery, patient empowerment and improvement of their quality of life. [6, 7].

Traditionally, therapies are usually provided to patients during their period of hospitalization by physical and occupational therapists and consist in mechanical exercises conducted by the therapists. However, in the last decades, many changes have been introduced in the rehabilitation of post-stroke patients. On the one hand, increasingly, treatments extend in time beyond the period of hospitalization and extend in the space, beyond the hospital to the patient's home [8]. On the other hand, new agents are involved in treatments, health professionals (doctors, nurses) and non-health professionals (engineers, exercise professionals, carers and family). Most of these changes have been made possible thanks to the development of technology [9].

2. Technological devices for upper limb rehabilitation

In the last 10 years, there has been increasing interest in the use of different technological devices for upper limb (UL) rehabilitation generally [5, 9], and particularly hand rehabilitation for stroke patients [10]. These studies have approached the problem from different points of view: (1) on the one hand, by analysing the physiological and psychophysical characteristics of different devices [11], (2) on the other analysing the key aspects of design and usability [12] and (3) finally studying its effectiveness in therapy [13, 14]. According to Kuchinke [12], these technical devices can be organized into two big groups: (1) on the one hand, devices based on virtual reality (VR) and (2) on the other robotic glove-like devices (GDs).

One of the main advantage of VRs and serious games [15] is to promote task-oriented and repetitive movement training of motor skill while using a variety of stimulating environments and facilitates adherence to treatment in the long term [16]. These devices can be used at home and in most cases do not require special investment in therapeutic hardware because they can use game consumables existing at home such as Nintendo(R) Wii[1] [17, 18], Leapmotion[2] [19, 20] or Kinect sensor[3] [21, 22]. Although first systematic studies based in VRs indicate that there is insufficient evidence to determine its effectiveness compared to conventional therapies [8], more recent studies [13, 14, 23] offer moderate evidence on the benefits of VR for UL motor improvement. Most researchers agree that VRs work well as coadjuvant to complement more conventional therapies; however, further studies with larger samples are needed to identify most suitable type of VR systems, to determine if VR results are sustained in the long term and to define the most appropriate treatment frequency and intensity using VR systems in post-stroke patients.

On the other hand, robotic systems and glove-like devices that provide extrinsic feedback like kinaesthetic and/or tactile stimulation have stronger evidence in the literature that improve motion ability of post-stroke patients [10, 24, 25]. Most of the evidences about effectiveness of GDs are based in pilot studies with non-commercial prototypes [26–30], but nowadays, there are also several commercial glove-like devices that support hand rehabilitation therapies for these patients such as HandTutor® [31, 32], Music Glove [33–35], Rapael Smart Glove [36] or CyberTouch [16, 37]. The main disadvantages of GDs are price, availability, because they are not yet widespread, and in some case the difficulty of setup handling and ergonomics.

As far as we know, there is little evidence in the literature supporting commercial glove-like devices for hand rehabilitation. This chapter presents a randomized clinical study (RCS) to test HandTutor® System in patients with chronic brain damage (CBD). There are some promising studies that show positive results by applying the HandTutor® in different groups of patients with stroke and traumatic brain injury (TBI) [31, 32], but samples include only people who are in the acute or subacute disease or injury but do not include chronic patients. This may be due to the added difficulty of obtaining positive results in interventions aimed at this group, in addition to the characteristics of adaptability and usability of the device that it is also harder for this kind of patients. The present work focuses on hand rehabilitation for chronic post-stroke patients.

3. Experimental design

We have conducted a pilot study (PS) to test acceptance, usability and adaptability of HandTutor® device in patients with chronic brain damage (CBD). This work describes setup, study protocol and preliminary results.

3.1. Participants description

Eligible participants met the following inclusion criteria: (1) At least 18-year age, (2) diagnosed with acquired brain injury: stroke or traumatic brain injury (TBI) and (3) chronic brain damage (more than 24 months from injury). In the final sample, 18 participants aged between

[1]https://www.nintendo.es/Wii/Wii-94559.html.
[2]https://www.leapmotion.com/.
[3]http://www.xbox.com/es-ES/xbox-one/accessories/kinect-for-xbox-one.

30 and 75 years old, 28% of subjects included in the pilot study are diagnosed with TBI and the remaining 72% of stroke; of these, more than half (56%) have left hemiplegia. The time from injury time exceeds 24 months, reaching 61% of cases 5 years of evolution. All the subjects included in the study attend regularly to a direct care acquired brain injury centre.

3.2. Device description

HandTutor® is a task-oriented device consistent on an ergonomic wearable glove and a laptop with rehabilitation software to enable functional training of hand, wrist and fingers. There are different models to fit both hands (left and right) and different sizes. The system allows the realization of an intensive and repetitive training but, at the same time, is flexible and adaptable to different motor abilities of patients after suffering a neurological, traumatological or rheumatological injury. The software allows the therapist to obtain different types of measures and to customize treatments for different patients, adapting the exercises to their physical and cognitive impairments. The HandTutor® provides augmented feedback and allows the participation of the user in different games that require practising their motor skills to achieve the game objective. Game objectives are highly challenging for patients and promote the improvement of deteriorated skills.

3.3. Study protocol

A randomized clinical trial (RCT) has been conducted with an experimental group and a control group. Participants in the experimental group have been treated with HandTutor® technological device, combined with conventional occupational therapy (set of functional tasks aimed at the mobility of the upper limb in ADLs). The control group only received conventional occupational therapy. All participants in the experimental group attend two weekly sessions with HandTutor®. Both groups received a weekly session of conventional therapy. It is a longitudinal study with pre-post intervention assessment, in which each subject is his control.

This chapter describes the first phase of the RCT, consisting of a pilot study (PS) to test the acceptance, usability and adaptability of the device by patients. For the PS, 18 patients of the global group were selected. Each subject completed four sessions using HandTutor® in both hands and a weekly session of COT. Each session includes quantitative and qualitative evaluation. The former one includes pre-intervention, and post-intervention assessment evaluating passive and active joint range of fingers and wrist, the latter include patients' interviews and therapist's observations. During the session, participants receive immediate visual and sensory feedback about their performance during exercises.

Each session includes a pre-intervention assessment and a back, wrist and hand. At the beginning of the session, the therapist evaluated the passive and active joint range of all fingers and wrist (flexion and extension). After the session, patient and therapist reviewed the increased joint range achieved during therapy on the joints involved. The software allows analysing and comparing the minimum and maximum levels in each of the movements required by the exercise. Each session lasts 45 minutes and consists of two exercises that focus their activity in flexion and extension of wrist and fingers independently, reaction speed and accuracy of the selected motion to move some elements included in the exercise.

First exercise of the session consisted in score as many balls as possible in the basket situated at the left of the patient. Every ball came to the patient from his right side. The goal of the second exercise of the session was destroying cylindrical rocks that were going from the right side to a planet situated in the left side. In both exercises, none of the elements appeared at the same height. That is why the patient had to adjust the degrees of flexion and extension of wrist, fingers or both. The occupational therapist could modify the speed, number of balls and minimum and maximum of degrees to achieve the accomplishment.

In addition to the quantitative variables described above, the therapist evaluated with qualitative methodology through interviews and observation, the condition of the skin (redness in the contact area with the glove), increased muscle tone, pain, motivation and difficulty understanding the instructions, level of usability, applicability and functionality of the patient. During the intervention, the therapist verbally corrected offsets trunk and lower limbs, annotating associated reactions in the facial muscles.

4. Results and discussion

All the participants of the experimental group completed the pilot study (n = 18). **Table 1** shows the passive and active range of motion (ROM) of the preseason evaluation in fingers and wrist, divided by diagnostic (stroke vs. traumatic brain injury). Every data about ROM is shown in millimetres (average score). In the evaluation, it is noted that the hand of the participants with traumatic brain injury showed lower passive and active joints in all of the fingers (active: V: 9, IV: 10, III: 9, II: 8 and I: 10; passive: V and IV: 14, III: 11, II: 17 and I: 16), except in the wrist (stroke: active 8; passive 23 vs. traumatic brain injury: active 18; passive 20).

Participants with stroke show higher deficits in the flexion active of the first, second and fifth fingers (9, 5.6 and 5.6, respectively), while the extension appears more weakened in the second and third fingers (7 and 8.4, respectively). However, the participants with traumatic brain injury show higher deficit of flexion in the third finger and the extension in the second, fourth and fifth fingers.

In every session, exercises were configured with the same reaction speed and the same number of objects, to allow the participants to achieve the maximum number of hits. Some of them showed deficit of attention, which means that the speed and the increase of stimulations could decrease the final scores and the motivation of the intervention. In the case of the participants who show spasticity, this speed allows them to autorelax and control the hand between the stimulations. The length of exercises were modified according to the muscular and attentional fatigue of the participant, starting with 5 minutes and decreasing, in some cases, up to 3 minutes. All the participants reached the accuracy of movement calculated by the system, according to the preseason ROM evaluation. Also, all of them were allowed to work all of the primary movement range calculated in the evaluation.

	Stroke (average in mm)		Traumatic brain injury (average in mm)	
Range of motion (flexo-extension)	**Active**	**Passive**	**Active**	**Passive**
Wrist	8	23	18	20
Little	11	20.3	9	14
Ring	14.3	22.6	10	14
Middle	11	22.6	9	11
Index	10	22.6	8	17
Thumb	8.3	20.6	10	16
Active flexion deficit				
Wrist	9		2	
Little	5.6		0	
Ring	5		0	
Middle	4.3		2	
Index	5.6		0	
Thumb	9		1	
Active extension deficit				
Wrist	6		0	
Little	3.3		5	
Ring	3.3		4	
Middle	8.4		0	
Index	7		9	
Thumb	3.3		5	
Treatments sessions log				
Reaction speed	10		10	
Accuracy	Full		Full	
Time in seconds	240		240	
(half)	1		1	
Number of objects	Full		Full	
Primary ranger				

Table 1. Hand ROM evaluation pre-session and treatments sessions log.

At the beginning of the session, the occupational therapist explained the exercise to the participant and conducted a 1-minute test to check understanding. Only was necessary to provide additional verbal instruction to improve comprehension in the 11% of the cases.

Figures 1 and **2** show the ROM evaluation of the hand. In **Figure 1**, the active evaluation of flexion of wrist and extension of fingers is observed. **Figure 2** includes the graphic representation of the millimetres of active movements (in red colour) versus the passive ones (in blue colour) of two hands with left hemiplegia (1 and 2) and two hands of participants with traumatic brain injury (tetraparesis and predominance of affectation in the right hemibody).

Figures 3 and **4** display the functioning of the HandTutor® during the intervention. **Figure 3** shows the glove with the hand in flexo-extension, while **Figure 4** shows the assisted movement of the occupational therapist to obtain the higher ranges of flexion in a participant who shows rigidity and attentional issues. Besides, in the contralateral hand, it can be seen the associated reactions in the top member, which is not forming a part of the intervention. The hand replicates the movement that the occupational therapist is trying to get in the most affected member.

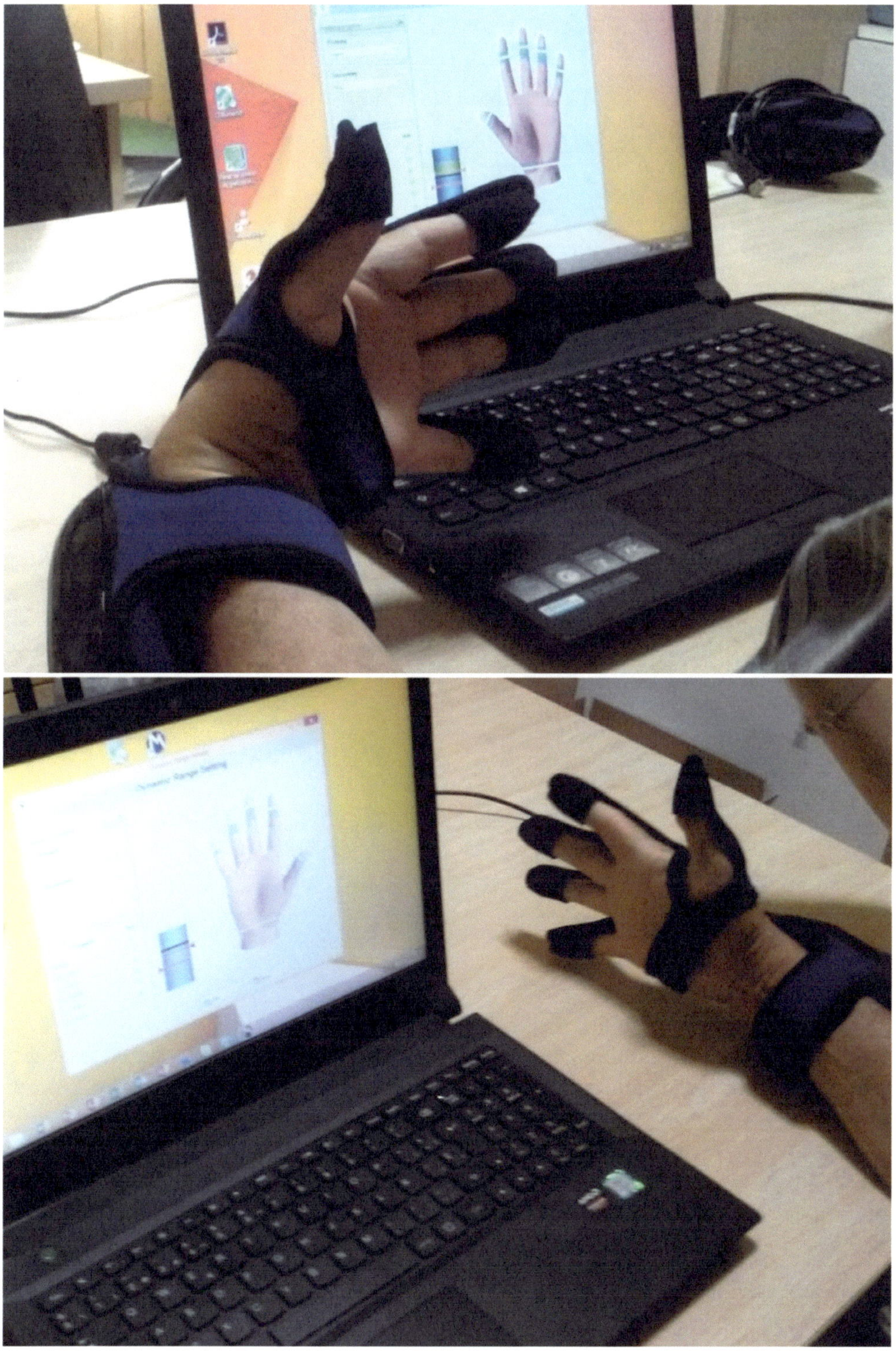

Figure 1. Hand ROM evaluation (active).

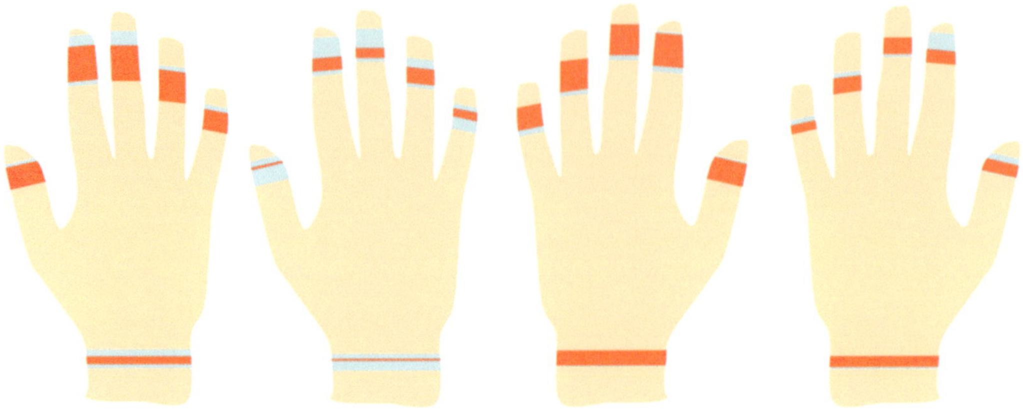

Figure 2. Hand ROM evaluation HandTutor® (passive and active).

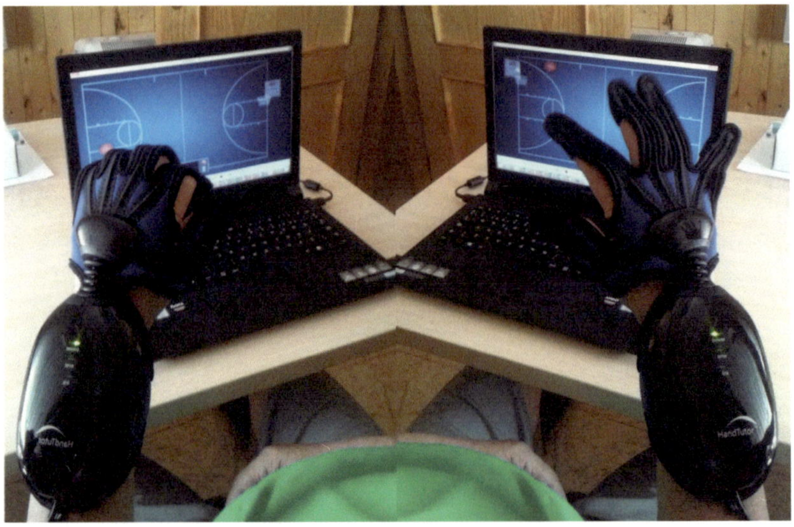

Figure 3. Flexo-extension hand with HandTutor®.

Figures 5 and **6** show maximum and minimum scores for diagnostic. In them, it can be observed the heterogeneity of the flexion and extension movement of the participants in the study. Regarding the wrist, it is not observed huge differences by diagnostic, except in the minimum flexo-extension of the stroke group, especially in the extension. Nevertheless, the articular ranges of the fingers differ until they reach a difference of 20 millimetres in the third finger in the case of the group diagnosed with stroke, coinciding with the group diagnosed with traumatic brain injury.

Participants referred increasing satisfaction with this new therapy. During the intervention, the software provided quantitative measures and immediate feedback of variations in patient mobility showing that HandTutor® sensors are highly sensitive to small variations in patient movement. In post-intervention interviews, patients reported that the glove decreases muscle tone of the hand and wrist, allowing ending the session with increased mobility.

http://dx.doi.org/10.5772/67532

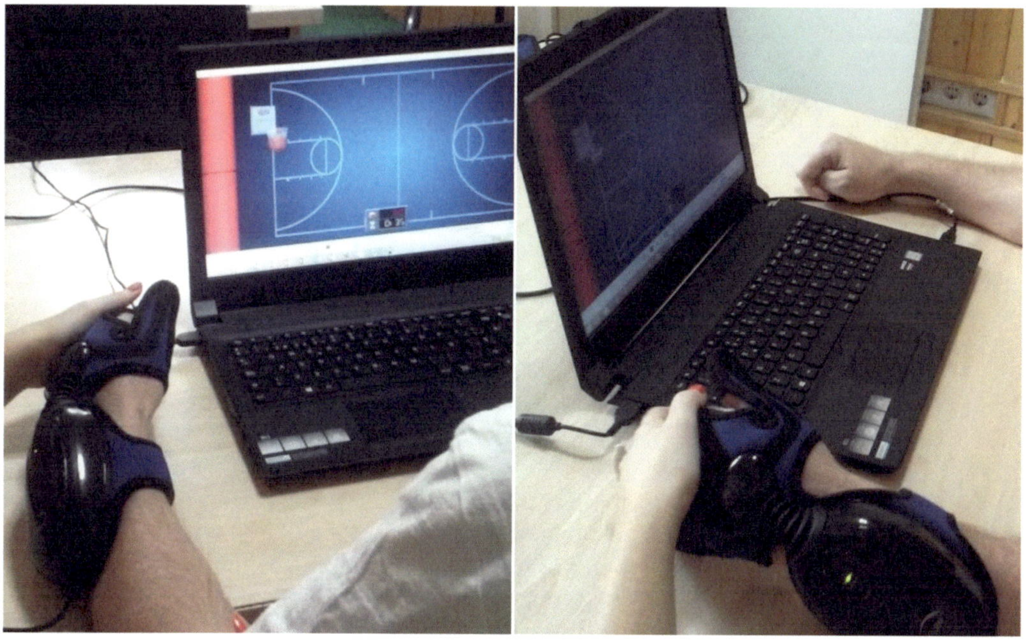

Figure 4. Example of assisted movement with HandTutor®.

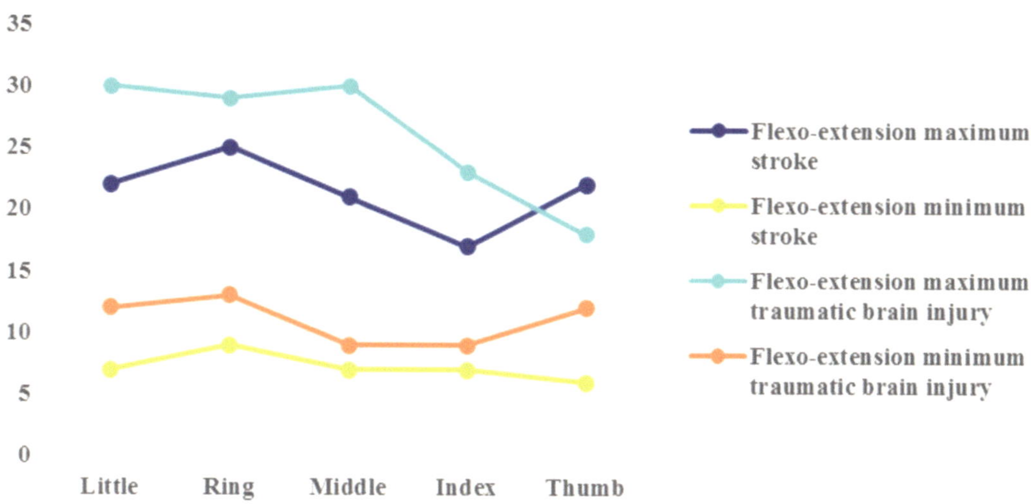

Figure 5. Flexo-extension maximum and minimum of fingers in treatments sessions log.

All sessions evaluated qualitatively, through an interview, the following parameters: skin condition, motivation, difficulty in the understanding of instructions, level of HandTutor® utility, clinic applicability and satisfaction.

During the sessions, no side effects were observed related to the skin or post-intervention pain related with the hand use. Every participant ended the sessions without any visible injury

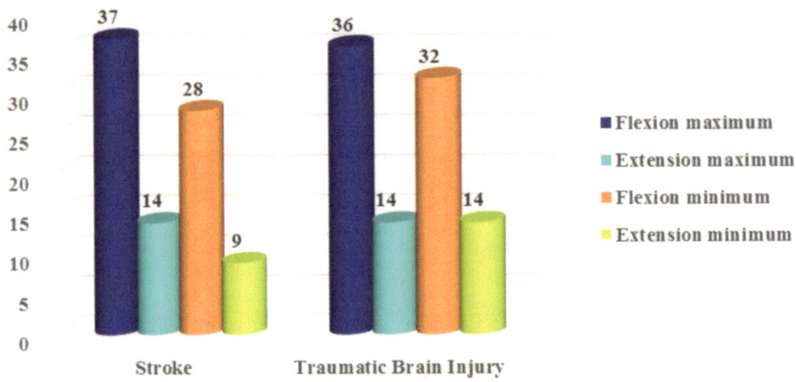

Figure 6. Flexo-extension maximum and minimum of wrist in treatments session logs.

in the skin (absence of redness, marks or changes in the coloration) and without any kind of pain. This was evaluated both at the end of the session and at the beginning of the next. To be able to contrast the information in relation with the skin condition and the pain, the data were triangulated by asking the participant and his/her primary caregiver the following day of every intervention. In both cases, they confirmed our data.

All participants referred high level of motivation and satisfaction at the end of the intervention due to the perceived higher performance of limb segments and joins involved in the exercises in their activities of daily life (ADLs). The subjective perception of the patient was checked by comparing the ROM (active vs. passive) pre-post measurement session. All participants showed and transmitted a great motivation and satisfaction with the HandTutor® intervention, except for one user. This one presents acoustic, visual and tactile hypersensitivity. After the pilot study, this participant transmitted that the glove, the sound and the images of the system induced in him/her nervousness and rejection. This information was contrasted with caregivers and professionals of the centre.

Some difficulties were found at the following of the exercise instructions, the motivation and interest maintenance during the 11% of the cases, as a consequence of the presence of attention and/or memory impairments.

All participants shared the sensation of decreasing the muscular tone, immediately at the end of every session and transmitted that this feeling stayed all day long, allowing them a higher mobility and independence at the ADLs.

During the study, some problems were observed associated with the difficulty in putting on the HandTutor® glove, especially in hands with high degrees of spasticity, mainly in diagnosed cases of traumatic brain injury (27.8%; **Figure 7**). Participants with lower ROM valued positively that the exercise was adapted to their possibilities, so they can reach and move objects even with their limited mobility. The 20% of the users valued negatively the weight of the system placed in the forearm, especially those with weak musculature. The occupational therapists reduced the gravity effect including a cradle to facility the placement of the forearm.

http://dx.doi.org/10.5772/67532

Figure 7. Spastic hand with HandTutor®.

In those patients that showed sweating, there were placed vinyl or latex gloves on their hands to avoid direct contact with the glove.

Therefore, it seems that the HandTutor® is a device with high degrees of acceptance and usability among patients with CBD.

5. Conclusions

This chapter is a step further of evaluating the acceptance of technological devices in chronic patients with CBD. On one hand, in the theoretical part of the study, we have found in the literature strong evidence confirming the effectiveness of glove-like devices in hand rehabilitation after brain injury, but no so solid evidence of VRs effectiveness over traditional treatment. On the other hand, the practical pilot study to test HandTutor points in the expected direction confirming participants' satisfaction about effectiveness and ergonomics of glove-like devices, but according to Ref. [12], there are still some issues to be solved in the usability of these devices for patients with spasticity.

The grade of usability of the HandTutor® device with chronic patients with CBD is high; we only find difficulties in those who show attention disorders and/or memory issues or sensorial hypersensibility. The degree of spasticity should also be taken into account in the design of the experience, because difficulties may arise in the placement of the device when the degree of spasticity is high or there is rigidity or other associated reactions.

Most of the studies performed with active gloves similar to HandTutor® device have been performed in patients in the acute or subacute phase of brain damage. It is important to emphasize that in this study, unlike the previous ones, the rehabilitation has been done with patients with more than 24 months of evolution since the diagnosis of the damage and therefore with a very high degree of chronicity in the neurological sequelae. This is one of the main contributions of the presented work since the more time has passed since the diagnosis of brain damage; the more difficult it is to achieve significant improvements with rehabilitation.

In our study, the HandTutor® device has performed effectively for the spasticity treatment in patients with CBD, producing improvements in the performance of the ADLs and elevating the motivation and satisfaction grades with his use in rehabilitation processes. However, this trial does not provide significant statistical evidence about HandTutor® effectiveness, and it would be recommendable to replicate the study with more participants to confirm our findings.

Acknowledgements

Work partially funded by SYMBHYO-ITC [MCINN PTQ-15-0705], RESET [MCINN TIN2014-53199-C3-1-R], eMadrid [CAM S2013/ICE-2715] and PhyMEL [UC3M 2015/00402/001] projects. Authors would like to thank Maria Pulido for their feedback in VR devices, and we also want to express our gratitude to the patients involved in the pilot study.

Author details

Marta Rodríguez-Hernández[1], Carmen Fernández-Panadero[2]*, Olga López-Martín[1] and Begoña Polonio-López[1]

*Address all correspondence to: mcfp@it.uc3m.es

1 University of Castilla La Mancha, Talavera de la Reina, Spain

2 University Carlos III, Madrid, Leganés, Spain

References

[1] Johnson, W., Onuma, O., Owolabic, M., & Sachdeva, S. (2016). Stroke: A global response is needed. Bulletin of the World Health Organization 94:634–634A. doi:10.2471/BLT.16.181636

[2] Feigin, V. L., Krishnamurthi, R. V., Parmar, P., Norrving, B., Mensah, G. A., Bennett, D. A., Barker-Collo S., Moran A.E., Sacco R.L., Truelsen T., Davis S., Pandian J.D., Naghavi M., Forouzanfar M.H., Nguyen G., Johnson C.O., Vos T., Meretoja A., Murray C.J.L. & Roth G.A. (2015). Update on the global burden of ischemic and hemorrhagic stroke in 1990–2013: The GBD 2013 study. Neuroepidemiology 45(3): 161–176. doi:10.1159/000441085

[3] Bleyenheuft, Y., & Gordon, A. M. (2014). Precision grip in congenital and acquired hemiparesis: Similarities in impairments and implications for neurorehabilitation. Frontiers in Human Neuroscience 8:459. doi:10.3389/fnhum.2014.00459

[4] Henderson, A., Korner-Bitensky, N., & Levin, M. (2014). Virtual reality in stroke rehabilitation: A systematic review of its effectiveness for upper limb motor recovery. Topics in Stroke Rehabilitation 14(2):52–61. doi:10.1310/tsr1402-52

[5] Hatem, S. M., Saussez, G., della Faille, M., Prist, V., Zhang, X., Dispa, D., & Bleyenheuft, Y. (2016). Rehabilitation of motor function after stroke: A multiple systematic review focused on techniques to stimulate upper extremity recovery. Frontiers in Human Neuroscience 10. doi:10.3389/fnhum.2016.00442

[6] Teasell, R. W., Foley, N. C., Bhogal, S. K., & Speechley, M. R. (2015). An evidence-based review of stroke rehabilitation. Topics in Stroke Rehabilitation 10(1):29–58. doi:10.1310/8YNA-1YHK-YMHB-XTE1

[7] Schneider, E. J., Lannin, N. A., Ada, L., & Schmidt, J. (2016). Increasing the amount of usual rehabilitation improves activity after stroke: A systematic review. Journal of Physiotherapy 62(4):182–187. doi:10.1016/j.jphys.2016.08.006

[8] Coupar, F., Pollock, A., Legg, L. A., Sackley, C., & van Vliet, P. (2012). Home based therapy programmes for upper limb functional recovery following stroke. The Cochrane Database Of Systematic Reviews 16;(5):CD006755. doi:10.1002/14651858. CD006755.pub2/abstract

[9] Pollock, A., Farmer, S. E., Brady, M. C., Langhorne, P., Mead, G. E., Mehrholz, J., & van Wijck, F. (2014). Interventions for improving upper limb function after stroke. The Cochrane Database Of Systematic Reviews 12;(11):CD010820. doi:10.1002/14651858

[10] Merrett, G. V., Metcalf, C. D., Zheng, D., Cunningham, S., Barrow, S., & Demain, S. H. (2011). Design and qualitative evaluation of tactile devices for stroke rehabilitation. In IET Seminar on Assisted Living, (pp. 1–6). IET London. http://eprints.soton.ac.uk/271802/1/merrett.pdf. doi:10.1049/ic.2011.0025

[11] Demain, S., Metcalf, C. D., Merrett, G. V., Zheng, D., & Cunningham, S. (2013). A narrative review on haptic devices: Relating the physiology and psychophysical properties of the hand to devices for rehabilitation in central nervous system disorders. Disability and Rehabilitation: Assistive Technology 8(3): 181–189. http://eprints.soton.ac.uk/339891/3/haptic_devices.pdf. doi:10.3109/17483107.2012.697532

[12] Kuchinke, L. M., & Bender, B. (2016). Technical view on requirements for future development of hand-held rehabilitation devices. In 6th IEEE International Conference on Biomedical Robotics and Biomechatronics (BioRob), (pp. 804–809). IEEE Singapore. doi:10.1109/BIOROB.2016.7523726

[13] Laver, K. E., George, S., Thomas, S., Deutsch, J. E., & Crotty, M. (2015). Virtual reality for stroke rehabilitation. The Cochrane Database Of Systematic Reviews 7;(9):CD008349. doi:10.1161/STROKEAHA.111.642439

[14] Viñas-Diz, S., & Sobrido-Prieto, M. (2016). Virtual reality for therapeutic purposes in stroke: A systematic review. Neurología (English Edition), 31(4): 255–277. doi:10.1016/j.nrleng.2015.06.007

[15] Joo, L. Y., Yin, T. S., Xu, D., Thia, E., Chia, P. F., Kuah, C. W. K., & He, K. K. (2010). A feasibility study using interactive commercial off-the-shelf computer gaming in upper limb

rehabilitation in patients after stroke. Journal of Rehabilitation Medicine 42(5): 437–441. doi:10.2340/16501977-0528

[16] Dimbwadyo-Terrer, I., Trincado-Alonso, F., de los Reyes-Guzmán, A., Aznar, M. A., Alcubilla, C., Pérez-Nombela, S., Ama-Espinosa, A., Polonio-López, B. & Gil-Agudo, Á. (2015). Upper limb rehabilitation after spinal cord injury: A treatment based on a data glove and an immersive virtual reality environment. Disability and Rehabilitation: Assistive Technology 11(6):462–467. doi:10.3109/17483107.2015.1027293

[17] Matamoros, M., Negrete, M., Haddad, Y., & Leder, R. S. (2010, March). Nintendo WII remote and Nunchuck as a wireless data subsystem for digital acquisition of analog physiologic data relevant to motor rehabilitation after stroke; part II. In 2010 Pan American Health Care Exchanges (pp. 198–200). Lima, Peru. IEEE. doi:10.1109/PAHCE.2010.5474568

[18] Adie, K., Schofield, C., Berrow, M., Wingham, J., Freeman, J., Humfryes, J., & Pritchard, C. (2014). Does the use of Nintendo Wii Sports™ improve arm function and is it acceptable to patients after stroke? Publication of the Protocol of the Trial of Wii™ in Stroke–TWIST. International Journal of General Medicine 7: 475. doi:10.2147/IJGM.S65379

[19] Eldem, C. (2014). Visual Neglect Assessment and Rehabilitation Using the Leap Motion. Saarbrücken: LAMBERT Academic Publishing

[20] Iosa, M. et al. (2015). Leap motion controlled videogame-based therapy for rehabilitation of elderly patients with subacute stroke: A feasibility pilot study. Topics in Stroke Rehabilitation 22(4): 306–316. doi:10.1179/1074935714Z.0000000036

[21] Shires, L., Battersby, S., Lewis, J., Brown, D., Sherkat, N., & Standen, P. (2013). Enhancing the tracking capabilities of the Microsoft Kinect for stroke rehabilitation. In IEEE 2nd International Conference on Serious Games and Applications for Health (SeGAH), (pp. 1–8). IEEE Portugal. doi:10.1109/SeGAH.2013.6665316

[22] Bamrungthai, P., & Pleehachinda, W. (2015, November). Development of a game-based system to support stroke rehabilitation using kinect device. In International Conference on Science and Technology (TICST), (pp. 323–326). IEEE London. doi:10.1109/TICST.2015.7369379

[23] Turolla, A., Dam, M., Ventura, L., Tonin, P., Agostini, M., Zucconi, C., & Piron, L. (2013). Virtual reality for the rehabilitation of the upper limb motor function after stroke: A prospective controlled trial. Journal of Neuroengineering and Rehabilitation 10(1): 1. doi:10.1186/1743-0003-10-85

[24] Lo, A. C., Guarino, P. D., Richards, L. G., Haselkorn, J. K., Wittenberg, G. F., Federman, D. G., Ringer R.J., Wagner T.H., Krebs H.I., Volpe B.T., Bever C.T.Jr., Bravata D.M., Duncan P.W., Corn B.H., Maffucci A.D., Nadeau S.E., Conroy S.S., Powell J.M., Huang G.D., & Peduzzi P. (2010). Robot-assisted therapy for long-term upper-limb impairment after stroke. New England Journal of Medicine 362(19): 1772–1783. doi:10.1056/NEJMoa0911341

[25] Subramanian, S. K., Massie, C. L., Malcolm, M. P., & Levin, M. F. (2010). Does provision of extrinsic feedback result in improved motor learning in the upper limb poststroke? A systematic review of the evidence. Neurorehabilitation and Neural Repair 24(2): 113–124. doi:10.1177/1545968309349941

[26] Huang, M. C., Xu, W., Su, Y., Lange, B., Chang, C. Y., & Sarrafzadeh, M. (2012). Smartglove for upper extremities rehabilitative gaming assessment. In Proceedings of the 5th International Conference on Pervasive Technologies Related to Assistive Environments (p. 20). ACM Heraklion, Crete, Greece. doi:10.1145/2413097.2413122

[27] Hiob, M. (2016). Interactive glove for mobility training and rehabilitation after stroke. Bachelor Thesis. In Certec - Rehabilitation Engineering and DesignCertec - Rehabilitation Engineering and DesignCertec - Rehabilitation Engineering and DesignCertec report TNS820 20161. Lund University Publications Student Papers. Lund, Sweden.

[28] Kang, B. B., Lee, H., In, H., Jeong, U., Chung, J., & Cho, K. J. (2016). Development of a polymer-based tendon-driven wearable robotic hand. In IEEE International Conference on Robotics and Automation (ICRA), 2016 (pp. 3750–3755). IEEE Stockholm. doi:10.1109/ICRA.2016.7487562

[29] Fischer, H. C., Triandafilou, K. M., Thielbar, K. O., Ochoa, J. M., Lazzaro, E. D., Pacholski, K. A., & Kamper, D. G. (2016). Use of a portable assistive glove to facilitate rehabilitation in stroke survivors with severe hand impairment. IEEE Transactions on Neural Systems and Rehabilitation Engineering 24(3): 344–351. doi:10.1109/TNSRE.2015.2513675

[30] Isnin, S. J., Hamid, D. H. T. A. H. & Sunar, S. Design of 3D immersive learning environment to enhanced hand skill after stroke. Man in India 96(6): 1727–1736.

[31] Carmeli, E., Vatine, J. J., Peleg, S., Bartur, G., & Elbo, E. (2009). Upper limb rehabilitation using augmented feedback: Impairment focused augmented feedback with HandTutor®. In 2009 Virtual Rehabilitation International Conference. Haifa, Israel. doi:10.1109/ICVR.2009.5174258

[32] Carmeli, E., Peleg, S., Bartur, G., Elbo, E., & Vatine, J. J. (2011). HandTutor® enhanced hand rehabilitation after stroke—A pilot study. Physiotherapy Research International 16(4): 191–200. doi:10.1002/pri.485

[33] Zondervan, D., Friedman, N. Chang, E., Zhao, X., & Cramer, S. C. (2016). Home-based hand rehabilitation after chronic stroke: Randomized, controlled single-blind trial comparing the MusicGlove with a conventional exercise program. Journal of Rehabilitation Research and Development 53(4): 457. doi:10.1682/JRRD.2015.04.0057

[34] Friedman, N., Chan, V., Zondervan, D., Bachman, M., & Reinkensmeyer, D. J. (2011, August). MusicGlove: Motivating and quantifying hand movement rehabilitation by using functional grips to play music. In 2011 Annual International Conference of the IEEE Engineering in Medicine and Biology Society (pp. 2359–2363). IEEE Boston, Massachusetts, USA doi:10.1109/IEMBS.2011.6090659

[35] Friedman, N., Chan, V., Reinkensmeyer, A. N., Beroukhim, A., Zambrano, G. J., Bachman, M., & Reinkensmeyer, D. J. (2014). Retraining and assessing hand movement after stroke using the MusicGlove: Comparison with conventional hand therapy and isometric grip training. Journal of Neuroengineering and Rehabilitation 11(1): 1. doi:10.1186/1743-0003-11-76

[36] Shin, J. H., Kim, M. Y., Lee, J. Y., Jeon, Y. J., Kim, S., Lee, S., Seo B., & Choi, Y. (2016). Effects of virtual reality-based rehabilitation on distal upper extremity function and health-related quality of life: A single-blinded, randomized controlled trial. Journal of Neuroengineering and Rehabilitation 13(1): 1. doi:10.1186/s12984-016-0125-x

[37] Kayyali, R., Shirmohammadi, S., El Saddik, A., & Lemaire, E. (2007). Daily-life exercises for haptic motor rehabilitation. In 2007 IEEE International Workshop on Haptic, Audio and Visual Environments and Games (pp. 118–123). IEEE Lyon, France.

Chapter 6

Impact of Non-Robotic Assisted Therapy for Improvement of Mobility of Paretic Upper Extremity Caused by Cerebral Palsy Compared to Classical Kinesiotherapy

Nina Sladekova, Elena Ziakova, Jaroslav Kresanek, Stanislava Klobucka, Jana Havlova and Miroslav Malay

Additional information is available at the end of the chapter

http://dx.doi.org/10.5772/67333

Abstract

Background: The aim of the clinical study was to investigate and compare the impact of non-robotic assisted therapy to classical kinesiotherapy to improve the function abilities of upper extremity.

Patients and methods: Sixty patients were divided randomly into two study groups. In the main group, patients completed a non-robotic assisted therapy and in the comparative group, they completed a classical kinesiotherapy. The age range of patients was from 6 to 17 years of age with impaired upper extremity. They all participated in 20 therapies.

Results: Statistically significant results were obtained in patients who completed the Armeo® therapy in all ranges of motion, the best improvement ($p = 0.000$) of shoulder and elbow flexion, and wrist extension, in all grips of the hand, the best improvement ($p = 0.000$) in lateral pinch, spherical and cylindrical grip and in Frenchay Arm Test in all tasks, the best improvement ($p = 0.000$) in tasks 1 and 5. The comparative group of the patients achieved statistically significant results only in elbow flexion ($p = 0.005$), radial deviation ($p = 0.046$), in ulnar deviation ($p = 0.011$). In other movements, grips and tasks were the results that are not statistically significant.

Conclusion: For the improvement of function ability of the paretic upper extremity, the patients with cerebral palsy are statistically more effective from the non-robotic assisted therapy than those who completed the classical kinesiotherapy.

Keywords: cerebral palsy, upper extremity, non-robotic assisted therapy, classical kinesiotherapy

---

1. Introduction

This clinical study has tested improvement of the movements of upper extremity in children and adolescents with cerebral palsy (CP). Arm rehabilitation is applied in neurorehabilitation for patients with paralyzed upper extremities due to lesions of the central or peripheral nervous system, for example, after stroke or spinal cord injury [1]. Lengthy physical inactivity in patients with chronic neurological disease can lead to prolonged recovery [2]. The goals of the therapy are to recover motor function, to improve movement coordination, to learn new motion strategies ("trick movements"), and/or to prevent secondary complications such as muscle atrophy, osteoporosis, and spasticity. The advantages of robotic training are that the therapist can get assisted, for example, relieved from the weight of the patient's arm, the training can get longer and more intensive (up to 20 times more movement repetitions per training session), and the movements can be measured and used for therapy assessment. Furthermore, special virtual reality technologies can make the training much more entertaining and motivating as well as task-oriented and functional and, thus, more relevant for daily living activities [1]. Cerebral palsy (CP) is defined as a group of permanent disorders of movement and posture, causing activity limitations attributed to a static lesion in the developing brain, often accompanied by secondary impairments. Predominant clinical manifestations found in CP include weakness, loss of selective motor control, spasticity, and antagonist contraction. Significant impairments caused by this disorder may compromise motor function, and as a result, individuals with CP experience functional limitations that affect activities of daily life ranging from mild incoordination to total body involvement [3]. One of the clinical features of cerebral palsy that perhaps has been least appreciated is impaired selective motor control (SMC). The National Institutes of Health Task Force defined SMC as the 'ability to isolate the activation of muscles in a selected pattern in response to demands of a voluntary movement or posture'. This can be extended to include movement of intended body segments in isolation. The intricate process of developing motor pathways establishing connections at the spinal-segmental level is susceptible to prenatal and perinatal brain damage that affect SMC. For example, it has long been established that the corticospinal tract directly innervates hand motor neurons, which provides the capacity for selective upper extremity movement control, and that damage to these tracts impairs this control [4]. Children and adolescents with CP have decreased levels of physical activity compared with their peers without CP. The ability to sustain physical activity at the intensity and duration necessary for participation is an important outcome of intervention. Young children with CP may be at risk for reduced physical activity and/or ability to sustain physical activity secondary to impairments in muscle performance, limitations in mobility, high calorie demands for growth, and decreased aerobic capacity [5]. Hemiparesis is usually a lifelong health problem, but is not unsolvable. By the effort to stifle debilitating disorder in hemiparesis and to therefore prevent its progression, it needs to be followed by restoration of lost functions and paretic upper extremity, which have created different methodological techniques and concepts. These are mostly based on the neurophysiologic basis [6].

1.1. Non-robotic therapy by Armeo® equipment

The therapy was implemented by means of the equipment Armeo®. The Armeo® equipment is an arm orthosis equipped with various components, including a pressure-sensitive handgrip. A spring mechanism provides adjustable weight support for the arm requiring treatment, which also facilitates functional arm movement. The Armeo® is used to support functional therapy for patients who lose function in their upper extremity caused by cerebral, neurogenic, spinal, muscular or bone-related disorders. Taking into account the contraindications and every patient's individual profile, the Armeo® is used in the case of: strokes, multiple sclerosis (MS), cerebral palsy (CP), follow-up care after brain-tumor operations, spinal cord injuries (SCI), traumatic brain injury (TBI), endoprostheses, follow-up care for elbow and shoulder endoprostheses, muscular atrophy, muscle weakness due to lack of mobility, hemiplegic patients. Just as for any other therapy, the physician in charge is always responsible for the indication. Functional training with the Armeo® is not possible or indicated in every case. In general, the Armeo® must not be used in the following cases to avoid causing harm to the patient. The following contraindications must therefore be observed in particular: orthosis cannot be fitted to the relevant arm, bone instability (non-consolidated fractures, severe osteoporosis), pronounced, fixed contractures affecting the relevant extremity, open skin lesions in the area of the relevant upper extremity, paraesthesia, shoulder joint subluxation or pain in the shoulder joint, severe spasticity, severe spontaneous movements, for example, ataxia, dyskinesia, myoclonic jerks, non-stable vital functions: pulmonary or cardio-circulatory contraindications (instability or instrumental support for these functions), need for long-term infusion therapy, severe postural instability, contraindicated sitting position, confused or non-cooperative patients, severe cognitive deficits, patients requiring isolation due to infections, severe visual problems (patient is not able to see displayed elements on the computer screen).

The Armeo® is based on the product "T-WREX". It is a passive (non-robotic) upper extremity orthosis, which lightens the weight of the upper extremity in 3D space. It allows natural movement in the workspace of approximately 66% of normal working area in the vertical and 72% in the horizontal plane. It allows quantifying range of motion and gripping strength in the patient's interaction with the software during therapy. This facilitates for users with moderate to severe hemiparesis to achieve greater range of motion that is possible without derating weight of the upper extremity. It also allows the use of upper extremity targeted and coordinated, although it retained residual possibility of movement. Since this is non-robotic, equipment requires the initiation of patient motion, which requires the active participation of the patient during training [7] (**Figures 1** and **2**).

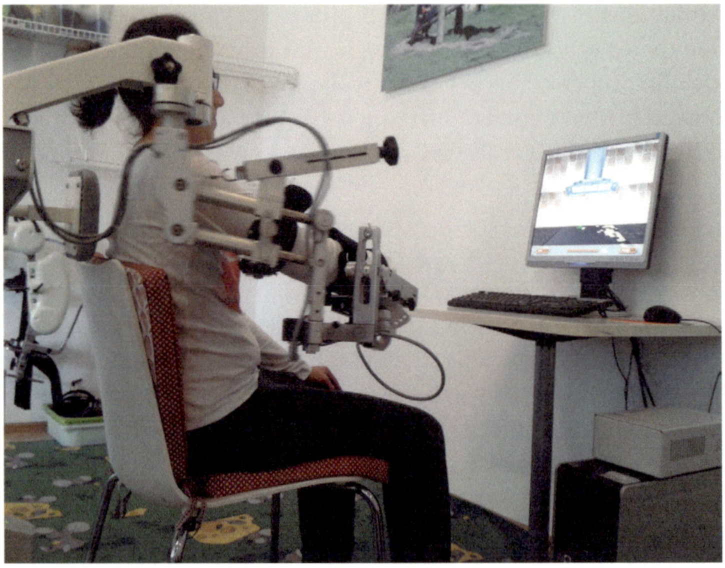

Figure 1. Therapy by using Armeo® equipment in 3D workspace.

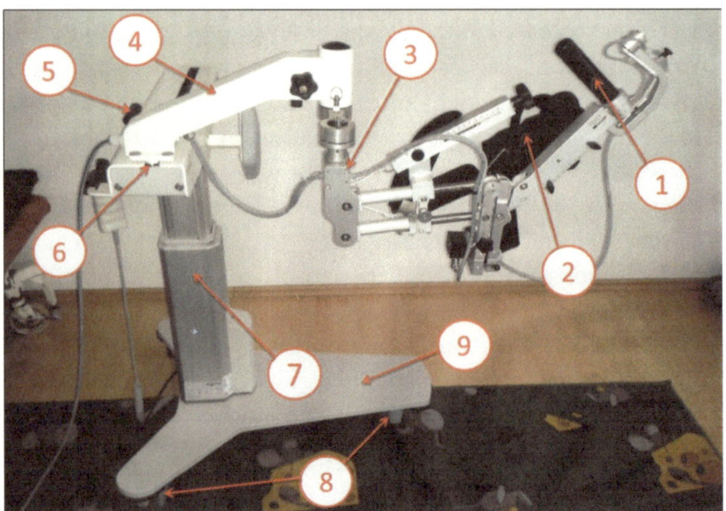

1.Caster
2.Base plate
3.Lifting column
4.Linear guide
5.Transport handle
6.Arm with integrated electronics
7.Upper arm module with weight compensation mechanism and cuff
8.Forearm module with weight compensation mechanism and cuff
9.Pressure sensitive handle with degree of freedom for pronation / supination

Figure 2. General overview of Armeo® equipment according to Hocoma (2008).

http://dx.doi.org/10.5772/67333

After the setting of therapy, the patient performs the specified sequence of exercise as an individual training. All exercises are performed in environment of virtual reality, which clearly displays functional tasks and performances of the patient [7]. Therapists can choose exercises, which they want to add to the users' therapy plan (e.g., Window Mopping, Reveal Panorama, Popping Air Bubbles) (**Figures 3** and **4**) [7]. Upon adding new exercises to the therapy plan, the plan definition screen appears. In this screen, all the exercise parameters such as difficulty level, time limits, number of repetitions and so on can be adjusted to the patients' needs. The Augmented Performance Feedback provided by the shared software platform, encourages and motivates patients to achieve a higher number of repetitions, and this leads to better, faster results and improved long-term outcomes. The software also provides automatic, ongoing assessment of motor functions and patients that can readily track their progress, helping them to grasp the initiative and reach toward recovery [7].

Figure 3. Grating carrot.

Figure 4. Shopping in 3D workplace.

The purpose of this clinical study was to determine the effect of therapy in the system Armeo® and on the movements and the grip's of the ability of upper extremity in children and adolescents with cerebral palsy. In this study, we sought to identify and verify the comparison of the impact of non-robotic assisted therapy to classical kinesiotherapy on the functionality effect of self-sufficiency and improvements of paretic upper extremity in the patients with CP. Even though we know that the complete elimination of paresis is impossible, we believe that paresis of the upper extremity can effect to a large extent, so that children and adolescents can improve their independence and quality of life.

2. Patients and methods

The object of investigation consisted of two groups. In the main group, patients completed a non-robotic assisted therapy and in the comparative group they completed a classical kinesiotherapy (e.g., passive movements, active-assisted exercises, Bobath concept, Kabat method). The age range of patients was from 6 to 17 years of age with impaired upper extremity. In the main group: 30 children (mean age 12.73) and in the comparative group: 30 children (mean age 11.33).They all have taken 20 therapies, whereas in the main group by Armeo® equipment and in the comparison group by classical kinesiotherapy. One therapy lasted 45 min of active exercise and frequency was minimal to twice a week. The patients were tested before and after the completion of therapy using goniometric investigation [7], by testing grips of paretic's hand (cylindrical, spherical, lateral, hook…) [8] and by using Frenchay Arm Test [9] for investigation of function ability of paretic upper extremity.

2.1. The statistical methods used in the study

For processing the collected data, a numerical evaluation and statistical methods were chosen. It was used as a descriptive analysis, Student's paired dependent t-test, Wilcoxon Signed Ranks Test and Effect size. Student's paired dependent t-test was used to evaluate the range of motion of the upper extremity for shoulder flexion and extension, wrist flexion and for evaluating the hand grip for palmar pinch, pinch grip and hook grip only. This test investigates the differences of two quantitative variables in the same investigating population. The result of the test is the t value (positive or negative), and significance. If the significance of the test is on the value higher than 0.05, then our observation of an intervention is not random. For other range of motions of the upper extremity and for other evaluation, the hand grip was used the Wilcoxon Signed Ranks Test – nonparametric statistical test, because in comparing to the test of the range of motions didn't work the test of normality for variances. This test does not compare the obtained values but the order of assigned values from the smallest to the largest. The study also shows the effect size. Effect size is used to obtain the size of standard rates of our observations. Effect size with significance, gives us information about the size and significance of the effect. Data were processed by using the software Microsoft Office Word 2007, Microsoft Office Excel, 2007. For mathematical-statistical evaluation, descriptive statistical methods SPSS 16.0 were used.

The study was conducted in accordance with ethical principles, based on the Declaration of Helsinki (1964) [10].

3. Results

In the main group after rehabilitation by equipment Armeo®, the patients achieved greater range of motions in the upper extremity than the patients in comparison group. After the testing of obtained input and output data, we used tests of normality (Kolmogorov-Smirnov and Shapiro-Wilk). The tests have confirmed homogeneous and inhomogeneous distribution of the data in the study, we used parametric statistical test—Student's paired dependent t-test and nonparametric statistical test—Wilcoxon Signed Ranks Test.

After the treatment had occurred in the patients of the main group, there was statistically significant improvements in range of motions of the upper extremity, which resulted in a higher average output score in shoulder flexion (M = 131.83, SD ± 29.55) than the input score (M = 111.33, SD ± 32.59), t(29) = −7.894, p = 0.000, r = 0.826, a higher average output score in shoulder abduction (Md = 100.00, SD ± 15.60) than the input score (Md = 82.50, SD ± 20.91), T = 0.00, Z(30) = −4.642, p = 0.000, r = −0.599, a higher average output score in elbow flexion (Md = 130.00, SD ± 11.84) than the input score (Md = 120.00, SD ± 13.61), T = 0.00, Z(30) = −4.342, p = 0.000, r = −0.561, a lower average output score in elbow extension (Md = 0.00, SD ± 4.69) than the input score (Md = 5.00, SD ± 7.03), T = 0.00, Z(30) = −3.397, p = 0.001, r = −0.439, because in elbow extension it has achieved the reduction until to elimination of flexion contractures, a higher average output score in wrist extension (Md = 30.00, SD ± 18.02) than the input score (Md = 20.00, SD ± 15.83), T = 0.00, Z(30) = −4.371, p = 0.000, r = −0.564, a higher average output score in radial deviation (Md = 20.00, SD ± 7.93) than the input score (Md = 20.00, SD ± 10.97), T = 0.00, Z(30) = −3.154, p = 0.002, r = −0.407. Statistically significant improvements also occurred in the other range of motions of the upper extremity (**Table 1**).

	Count	Mean	Maximum	Minimum	Median	Standard error of mean	Standard deviation	Student t-test/ Wilcoxon signed ranks test	Sig. (two-tailed)	Effect size
Shoulder flex. input	30	111.33	170	30	110	5.950	32.588	t = −7.894	0.000	r = 0.826
Shoulder flex. output	30	131.83	180	60	130	5.395	29.551			
Shoulder ext. input	30	23.83	50	10	20	2.001	10.961	Z = −3.823	0.000	r = −0.493
Shoulder ext. output	30	29.83	50	20	30	1.864	10.212			
Shoulder abd. input	30	83.83	130	20	82.50	3.818	20.914	Z = −4.642	0.000	r = −0.599
Shoulder abd. output	30	98.50	130	70	100	2.848	15.600			
Shoulder add. input	30	15.17	25	0	20	1.085	5.943	Z = −2.972	0.003	r = −0.384
Shoulder add. output	30	18.17	30	10	20	1.002	5.490			
Elbow flex. input	30	118.17	150	90	120	2.485	13.613	Z = −4.342	0.000	r = −0.561
Elbow flex. output	30	130.50	150	110	130	2.162	11.843			
Elbow ext. input	30	6.17	20	0	5	1.284	7.032	Z = −3.397	0.001	r = −0.439
Elbow ext. output	30	2.50	20	0	0	0.856	4.689			

http://dx.doi.org/10.5772/67333

	Count	Mean	Maximum	Minimum	Median	Standard error of mean	Standard deviation	Student t-test/ Wilcoxon signed ranks test	Sig. (two-tailed)	Effect size
Wrist ext. input	30	20.50	50	0	20	2.890	15.830	Z = −4.371	0.000	r = −0.564
Wrist ext. output	30	31.67	60	5	30	3.290	18.020			
Wrist flex. input	30	67.50	110	30	70	4.088	22.390	t = −6.456	0.000	r = 0.768
Wrist flex. output	30	76.67	130	50	80	3.670	20.100			
Radial deviat. input	30	15.67	50	0	20	2.002	10.965	Z = −3.154	0.002	r = −0.407
Radial deviat. output	30	20.17	35	5	20	1.448	7.931			
Ulnar deviat. input	30	17.33	30	0	20	1.413	7.739	Z = −4.288	0.000	r = −0.554
Ulnar deviat. output	30	24.67	40	10	22.50	1.290	7.063			

Flex., flexion; ext., extension; abd., abduction; add., adduction; deviat., deviation.

Table 1. Descriptive statistic of the measurement range of motion of the upper extremity in the main group of patients, who completed non-robotic therapy.

In the patients of the comparison group, there was statistically significant improvements in range of motions of the upper extremity only in the three motions, which resulted in a higher average output score in elbow flexion (Md = 140.00, SD ± 13.88) than the input score (Md = 140.00, SD ± 13.88), T = 0.00, Z(30) = −2.828, p = 0.005, r = −0.365, a higher average output score in radial deviation (Md= 10.00, SD ± 9.21) than the input score (Md = 10.00, SD ± 9.26), T = 0.00, Z(30) = −2.000, p = 0.046, r = −0.258, a higher average output score in ulnar deviation (Md = 20.00, SD ± 14.37) than the input score (Md = 20.00, SD ± 13.81), T = 0.00, Z(30) = −2.530, p = 0.011, r = −0.327. In the other range of motions have not occurred statistically significant improvements of the upper extremity (**Table 2**).

	Count	Mean	Maximum	Minimum	Median	Standard error of mean	Standard deviation	Student t-test/ Wilcoxon signed ranks test	Sig. (two-tailed)	Effect size
Shoulder flex. input	30	99.50	150	60	90	4.357	23.866	Z = −1.633	0.102	*
Shoulder flex. output	30	100.23	160	60	90	4.447	24.359			
Shoulder ext. input	30	27.67	60	0	30	2.783	15.241	t = −1.439	0.161	*
Shoulder ext. output	30	28.00	60	0	30	2.729	14.948			
Shoulder abd. input	30	85.50	130	20	90	3.493	19.134	Z = −1.000	0.317	*
Shoulder abd. output	30	85.67	130	20	90	3.505	19.197			
Shoulder add. input	30	26.83	45	0	22.50	2.790	15.284	Z = −1.414	0.157	*
Shoulder add. output	30	27.17	45	5	25	2.741	15.011			
Elbow flex. input	30	132.5	145	100	140	2.534	13.881	Z = −2.828	0.005	r = −0.365
Elbow flex. output	30	133.83	150	100	140	2.533	13.877			
Elbow ext. input	30	7.83	40	0	5	1.773	9.710	Z = −1.414	0.157	*
Elbow ext. output	30	7.50	40	0	5	1.741	9.537			
Wrist ext. input	30	32.50	70	0	30	4.233	23.184	Z = 0.000	1.000	*
Wrist ext. output	30	32.50	70	0	30	4.233	23.184			
Wrist flex. input	30	42.17	60	10	47.50	3.039	16.645	t = −1.633	0.102	*
Wrist flex. output	30	42.83	60	10	47.50	2.982	16.331			

	Count	Mean	Maximum	Minimum	Median	Standard error of mean	Standard deviation	Student t-test/ Wilcoxon signed ranks test	Sig. (two-tailed)	Effect size
Radial deviat. input	30	12.83	30	0	10	1.690	9.255	Z = −2.000	0.046	r = −0.258
Radial deviat. output	30	13.50	30	0	10	1.681	9.206			
Ulnar deviat. input	30	21.17	50	0	20	2.522	13.814	Z = −2.530	0.011	r = −0.327
Ulnar deviat. output	30	22.50	55	0	20	2.623	14.369			

Flex., flexion; ext., extension; abd., abduction; add., adduction; deviat., deviation.
*Not-statistically significant $p \geq 0.05$.

Table 2. Descriptive statistic of the measurement range of motion of the upper extremity in the comparison group of patients, who completed classical kinesiotherapy.

Significantly better results demonstrated the improvement in hand grip in the main group of patients, which resulted in a higher average output score in lateral pinch (Md = 3.00, SD ± 1.19) compared with the input score (Md = 2.00, SD ± 1.44), T = 0.00, Z(30) = −4.264, p = 0.000, r = −0.550, a higher average output score in spherical grip (Md = 4.00, SD ± 0.92) compared with the input score (Md = 3.00, SD ± 1.30), T = 0.00, Z(30) = −4.400, p = 0.000, r = −0.568, a higher average output score in cylindrical grip (Md = 3.50, SD ± 1.22) compared with the input score (Md = 2.00, SD ± 1.32), T = 0.00, Z(30) = −4.534, p = 0.000, r = −0.589, a higher average output score in key (lateral) grip (Md = 2.00, SD ± 1.22) compared with the input score (Md = 1.00, SD ± 1.31), T = 0.00, Z(30) = −4.001, p = 0.000, r = −0.516, a higher average output score in scissors grip (Md = 1.00, SD ± 1.45) compared with the input score (Md = 0.00, SD ± 1.39), T = 0.00, Z(30) = −4.000, p = 0.000, r = −0.516, a higher average output score in conical grip (Md = 3.00, SD ± 1.22) compared with the input score (Md = 2.00, SD ± 1.37), T = 0.00, Z(30) = −4.025, p = 0.000, r = −0.520 (**Table 3**).

	Count	Mean	Maximum	Minimum	Median	Standard error of mean	Standard deviation	Wilcoxon signed ranks test	Asymp. sig. (two-tailed)	Effect size
Palmar pinch input	30	1.90	5	0	1.50	0.232	1.269	Z = −3.771	0.000	r = −0.487
Palmar pinch output	30	2.43	5	1	2.00	0.238	1.305			

	Count	Mean	Maximum	Minimum	Median	Standard error of mean	Standard deviation	Wilcoxon signed ranks test	Asymp. sig. (two-tailed)	Effect size
Tip to tip pinch input	30	1.97	5	0	2.00	0.277	1.520	Z = −3.742	0.000	r = −0.483
Tip to tip pinch output	30	2.43	5	0	2.00	0.290	1.591			
Pinch grip input	30	1.60	5	0	1.00	0.233	1.276	Z = −3.742	0.000	r = −0.483
Pinch grip output	30	2.07	5	0	2.00	0.230	1.258			
Tabletop grip input	30	1.17	4	0	1.00	0.215	1.177	Z = −3.162	0.002	r = −0.408
Tabletop grip output	30	1.50	5	0	1.00	0.248	1.358			
Lateral pinch input	30	2.30	5	0	2.00	0.263	1.442	Z = −4.264	0.000	r = −0.550
Lateral pinch output	30	2.97	5	1	3.00	0.217	1.189			
Spherical grip input	30	3.03	5	1	3.00	0.237	1.299	Z = −4.400	0.000	r = −0.568
Spherical grip output	30	3.90	5	2	4.00	0.168	0.923			
Cylindrical grip input	30	2.67	5	0	2.00	0.241	1.322	Z = −4.534	0.000	r = −0.589
Cylindrical grip output	30	3.53	5	1	3.50	0.224	1.224			
Hook grip input	30	1.30	4	0	1.00	0.221	1.208	Z = −2.972	0.003	r = −0.384
Hook grip output	30	1.70	4	0	1.00	0.254	1.393			
Claw grip input	30	1.43	4	0	1.00	0.218	1.194	Z = −3.500	0.000	r = −0.452
Claw grip output	30	1.90	5	0	1.50	0.26	1.423			
Tip to palm distance input	30	0.60	7	0	0.00	0.286	1.567	Z = −2.032	0.042	r = −0.262
Tip to palm distance output	30	0.23	4	0	0.00	0.141	0.774			
Key (lateral) grip input	30	1.50	5	0	1.00	0.239	1.306	Z = −4.001	0.000	r = −0.516
Key (lateral) grip output	30	2.20	5	1	2.00	0.222	1.215			

	Count	Mean	Maximum	Minimum	Median	Standard error of mean	Standard deviation	Wilcoxon signed ranks test	Asymp. sig. (two-tailed)	Effect size
Pencil grip input	30	1.87	5	0	1.00	0.261	1.432	Z = −3.494	0.000	r = −0.451
Pencil grip output	30	2.43	5	1	2.00	0.228	1.251			
Tweezers grip input	30	1.30	5	0	1.00	0.250	1.368	Z = −3.317	0.001	r = −0.428
Tweezers grip output	30	1.67	5	0	1.50	0.251	1.373			
Scissors grip input	30	1.07	5	0	0.00	0.253	1.388	Z = −4.000	0.000	r = −0.516
Scissors grip output	30	1.60	5	0	1.00	0.265	1.453			
Conical grip input	30	2.00	5	0	2.00	0.249	1.365	Z = −4.025	0.000	r = −0.520
Conical grip output	30	2.60	5	0	3.00	0.223	1.221			

Table 3. Descriptive statistics of the testing grips of paretic's hand in the main group of patients, who completed non-robotic therapy.

In testing grips of paretic's hands, there were statistically significant results of the main group of patients, unlike of the comparative group of patients, where they have not achieved statistically significant results in testing of paretic's hands (**Table 4**).

	Count	Mean	Maximum	Minimum	Median	Standard error of mean	Standard deviation	Student t-test/ Wilcoxon signed ranks test	Asymp. sig. (two-tailed)	Effect size
Palmar pinch input	30	2.87	5	0	3.00	0.313	1.717	t = −1.795	0.083	*
Palmar pinch output	30	2.97	5	0	3.00	0.294	1.608			
Tip to tip pinch input	30	2.80	5	0	3.00	0.337	1.846	Z = −1.732	0.083	*
Tip to tip pinch output	30	2.90	5	0	3.00	0.312	1.709			
Pinch grip input	30	2.37	5	0	2.00	0.327	1.790	t = −1.439	0.161	*
Pinch grip output	30	2.43	5	0	2.50	0.321	1.755			

	Count	Mean	Maximum	Minimum	Median	Standard error of mean	Standard deviation	Student t-test/ Wilcoxon signed ranks test	Asymp. sig. (two-tailed)	Effect size
Tabletop grip input	30	1.93	5	0	2.00	0.321	1.760	Z = −1.414	0.157	*
Tabletop grip output	30	2.00	5	0	2.00	0.314	1.722			
Lateral pinch input	30	2.57	5	0	3.00	0.313	1.716	Z = −1.000	0.317	*
Lateral pinch output	30	2.60	5	0	3.00	0.306	1.673			
Spherical grip input	30	3.40	5	0	4.00	0.286	1.567	Z = −1.414	0.157	*
Spherical grip output	30	3.47	5	0	4.00	0.278	1.525			
Cylindrical grip input	30	3.63	5	0	4.00	0.297	1.629	Z = −1.732	0.083	*
Cylindrical grip output	30	3.73	5	0	4.00	0.291	1.596			
Hook grip input	30	2.43	5	0	2.50	0.298	1.633	t = −1.439	0.161	*
Hook grip output	30	2.50	5	0	2.50	0.295	1.614			
Claw grip input	30	2.13	5	0	2.00	0.310	1.697	Z = −0.000	1.000	*
Claw grip output	30	2.13	5	0	2.00	0.310	1.697			
Tip to palm distance input	30	0.53	5	0	0.00	0.243	1.332	Z = −1.414	0.157	*
Tip to palm distance output	30	0.47	5	0	0.00	0.218	1.196			
Key (lateral) grip input	30	2.97	5	0	3.00	0.320	1.752	Z = −1.414	0.157	*
Key (lateral) grip output	30	3.03	5	0	3.00	0.301	1.650			
Pencil grip input	30	2.50	5	0	3.00	0.306	1.676	Z = −1.732	0.083	*
Pencil grip output	30	2.60	5	0	3.00	0.309	1.694			

	Count	Mean	Maximum	Minimum	Median	Standard error of mean	Standard deviation	Student t-test/ Wilcoxon signed ranks test	Asymp. sig. (two-tailed)	Effect size
Tweezers grip input	30	2.47	5	0	2.00	0.348	1.907	Z = −1.000	0.317	*
Tweezers grip Output	30	2.50	5	0	2.00	0.342	1.871			
Scissors grip Input	30	2.43	5	0	2.50	0.317	1.736	Z = −0.000	1.000	*
Scissors grip output	30	2.43	5	0	2.50	0.317	1.736			
Conical grip input	30	2.87	5	0	3.00	0.338	1.852	Z = −0.000	1.000	*
Conical grip output	30	2.87	5	0	3.00	0.338	1.852			

*Not-statistically significant $p \geq 0.05$.

Table 4. Descriptive statistics of the testing grips of paretic's hand in the comparison group of patients, who completed classical kinesiotherapy.

By testing the Frenchay Arm Test, significant improvements have occurred in patients of the main group in all tasks, where the highest statistically significance was tasks 1 and 5, which resulted in a higher average output score in task 1 (Md = 1.00, SD ± 0.31) than the input score (Md = 0.00, SD ± 0.51), T = 0.00, Z(30) = −3.606, p = 0.000, r = −0.465, a higher average output score in task 5 (Md = 1.00, SD ± 0.35) compared with the input score (Md = 0.00, SD ± 0.47), T = 0.00, Z(30) = −4.123, p = 0.000, r = −0.532 (**Table 5**).

	Count	Mean	Maximum	Minimum	Median	Standard error of mean	Standard deviation	Wilcoxon signed ranks test	Asymp. sig. (two-tailed)	Effect size
Task 1 input	30	0.47	1	0	0.00	0.093	0.507	Z = −3.606	0.000	r = −0.465
Task 1 output	30	0.90	1	0	1.00	0.056	0.305			
Task 2 input	30	0.33	1	0	0.00	0.088	0.479	Z = −2.646	0.008	r = −0.342
Task 2 output	30	0.57	1	0	1.00	0.092	0.504			
Task 3 input	30	0.20	1	0	0.00	0.074	0.407	Z = −2.646	0.008	r = −0.342
Task 3 output	30	0.43	1	0	0.00	0.092	0.504			

	Count	Mean	Maximum	Minimum	Median	Standard error of mean	Standard deviation	Wilcoxon signed ranks test	Asymp. sig. (two-tailed)	Effect size
Task 4 input	30	0.13	1	0	0.00	0.063	0.346	Z = −2.236	0.025	r = −0.289
Task 4 output	30	0.30	1	0	0.00	0.085	0.466			
Task 5 input	30	0.30	1	0	0.00	0.085	0.466	Z = −4.123	0.000	r = −0.532
Task 5 input	30	0.87	1	0	1.00	0.063	0.346			

Table 5. Descriptive statistics of the testing Frenchay Arm Test in the main group of patients, who completed non-robotic therapy.

In the patients of the comparison group, improvements have not occurred of statistically significant in task of Frenchay Arm Test (**Table 6**).

	Count	Mean	Maximum	Minimum	Median	Standard error of mean	Standard deviation	Wilcoxon signed ranks test	Asymp. sig. (two-tailed)	Effect size
Task 1 input	30	0.67	1	0	1.00	0.088	0.479	Z = −1.732	0.083	*
Task 1 output	30	0.77	1	0	1.00	0.079	0.430			
Task 2 input	30	0.77	1	0	1.00	0.079	0.430	Z = −0.000	1.000	*
Task 2 output	30	0.77	1	0	1.00	0.079	0.430			
Task 3 input	30	0.43	1	0	0.00	0.092	0.504	Z = −0.000	1.000	*
Task 3 output	30	0.43	1	0	0.00	0.092	0.504			
Task 4 input	30	0.47	1	0	0.00	0.093	0.507	Z = −0.000	1.000	*
Task 4 output	30	0.47	1	0	0.00	0.093	0.507			
Task 5 input	30	0.57	1	0	1.00	0.092	0.504	Z = −0.000	1.000	*
Task 5 input	30	0.57	1	0	1.00	0.092	0.504			

*Not-statistically significant $p \geq 0.05$.

Table 6. Descriptive statistics of the testing Frenchay Arm Test in the comparison group of patients, who completed classical kinesiotherapy.

4. Discussion

Robot assisted upper extremity therapy has been shown to be effective in adult stroke patients and in children with cerebral palsy (CP) and other acquired brain injuries (ABI). The patient's active involvement is a factor with its effectiveness. However, this demands focused attention during training sessions, which can be a challenge for children [11]. We agree with the authors, however, with our children, we would like to highlight the increased attention needed, because then the games would interest them and they would be completely focused on the therapy. Krebs [12] published a study, where he tested in children with cerebral palsy (CP). He tested whether or not motor habilitation resembles motor learning. Twelve children with hemiplegic CP, aged 5 – 12 years with moderate to severe motor impairments underwent a 16-session robot-mediated planar therapy program to improve their upper extremity reach, with a focus on shoulder and elbow movements. Participants were trained to execute point-to-point movements (with robot assistance) with the affected arm and were evaluated (without robot assistance) in trained (point-to-point) and untrained (circle-drawing) conditions. Outcomes were measured at baseline, midpoint, immediately after the program, and 1-month post completion. Outcomes measured were the Fugl-Meyer (FM), Quality of Upper Extremity Skills Test (QUEST), and Modified Ashworth Scale (MAS) scores, parent questionnaire, and robot-based kinematic metrics. After robotic intervention, the authors found significant gains in the FM, QUEST, and parent questionnaire. Robot-based evaluations demonstrated significant improvement in trained movements and that improvement was sustained at follow-up. Furthermore, children improved their performance in untrained movements indicating generalization. Therapy in our study was focused to determine the effect of non-robotic assisted therapy for children with cerebral palsy. We focused on improving the range of motions in the upper extremity, improving grips of paretic hand and on testing of Frenchay Arm Test.

Robotic and non-robotic training devices are increasingly being used in the rehabilitation of upper extremity function in subjects with neurological disorders. As well as being used for training such devices can also provide ongoing assessments during the training sessions. Therefore, it is mandatory to understand the reliability and validity of such measurements when used in a clinical setting [13]. We consent, therefore, started using non-robotic Armeo® equipment in our rehabilitation center.

Lo and colleagues [14] demonstrated that the robotic system for shoulder/elbow rehabilitation on chronic post-stroke patients did not significantly improve motor performance after 12 weeks compared to usual care or intensive therapy. Nevertheless, secondary analyses showed that the robot-assisted therapy compared to usual care rather than intensive therapy improved outcomes over 36 weeks. We achieved in our clinical study statistically significant results after the completion of 20 therapies in non-robotic equipment of patients with cerebral palsy compared to the comparative group of patients who have completed classical kinesiotherapy.

Studies have confirmed significant improvement in mobility of the upper extremity in patients with hemiparesis. It has increased the muscle strength, increased the range of joint mobility, improved the neuromuscular coordination, improved the upper extremity function, and increased the patient's motivation and lastly the improvement of self-sufficiency. The results of the available studies have supported the current theory of motor learning by repeating the motions, which it describes the correlation between the repetition of activities and improving motor function, therefore being the key to stimulate motor plasticity [15]. Recent studies have demonstrated that robot-assisted therapy, in combination with new rehabilitation techniques, motivates the patient (which is very important in the case of children) and improves the treatment. A new and advanced method of feedback is the application of virtual scenarios, where the user can interact with a virtual object in real time and feels that he or she is part of a virtual environment during the therapy. Changes in cortical maps are driven by specific aspects of behavioral demand (i.e., motivation, skill acquisition) and are not simply the result of repetitive use or strength training. Virtual reality is a very attractive tool to enable the adoption of biofeedback techniques for the treatment of children with CP. In this scenario, biofeedback can be defined as the use of sensory feedback through which objective performance observation related to a specific motor task is presented to provide the child with immediate, consistent feedback of their performance. The aim of providing patients with biofeedback during exercise is twofold: first, to improve the effectiveness of the rehabilitation treatment, both by allowing patients to adjust their movements according to the feedback of performance and by providing an incentive to exercise; and second, recording the physiological parameters to be fed back to the patient, provides quantitative monitoring and documentation of the patient's progress during treatment. The latter feature is particularly important when the rehabilitation treatment is extensive and prolonged, which is typically the case with patients with CP [16]. We have to agree here with the authors of international clinical studies, because it has showed greater interest in the therapy from the patient's side and greater motivation especially in children and adolescence age, where it is well known that it is difficult to motivate and to improve attention in therapy. There is evidence that not only severe stressful events, but also common low-threat events, in particular chronic ones, may cause or provoke some mental disorders, especially in childhood [17]. Patient motivation is absolutely critical for successful rehabilitation after neurological injury. First, motivation in terms fun is important to maintain compliance, on a psychological level. Second, recent neuroscience research has shown that obtaining reward and challenge can enhance performance even on a deeper, neuro-physiological level [18]. A study in non-clinical populations demonstrated that depression diminishes the capability of imagining future positive outcomes and strengthens the ability to imagine negative outcomes. Patients with affective disorders also present cognitive dysfunction in areas such as working memory, attention and learning. Depression has been shown to significantly impair attention and word memory [19]. From our experience, child and adolescent patients with cerebral palsy are often depressed, especially when therapy is less effective or when progressing very slowly, we want to highlight the therapy by equipment Armeo® where we utilize motivation and cooperation of the patient, and therefore the therapy is more effective and faster.

The existing shortage of therapists and caregivers assisting physically disabled individuals at home is expected to increase and become serious problem in the near future. The patient population needing physical rehabilitation of the upper extremity is also constantly increasing. Robotic devices have the potential to address this problem as noted by the results of recent research studies. However, the availability of these devices in clinical settings is limited, leaving plenty of room for improvement [20]. Rehabilitation programs based on robotics adapted to the special needs of an individual user are expensive and therefore limited resources hinder the achievement of optimal therapy. Moreover, specialized technicians are needed to control the robotic technology, and this means higher costs to the family and society [16]. Despite the success of the treatment of non-robotic equipment, we are in our rehabilitation center, the only one who owns a non-robotic equipment of Armeo® in Slovak Republic.

5. Conclusion

This clinical study has achieved statistically significant results in the main group of the patients with cerebral palsy, who completed non-robotic assisted therapy compared to the comparative group of the patients who have completed classical kinesiotherapy. Therapy has improved the range of motions in the upper extremity; similarly, significant results have been shown in improvements in grip ability of paretic hand and by testing Frenchay Arm Test in the patients of the main group. The co-operation with patients during the non-robotic assisted therapy was very good. They were coming to the therapy regularly and really looking forward to it. We can say based on the analysis results, that non-robotic assisted therapy of Armeo® positively effects the rehabilitation of the children and adolescents with cerebral palsy. We would like to emphasize not only the positive effect of therapy, but also the patient's successfulness of motivation in the adolescent age. Although the therapy in system of Armeo® is more costly than conventional methods, successfulness of the treatment has a very high rate. As we know, we can never completely get a patient with cerebral palsy back to full health, but we can help them to improve the function abilities of paretic upper extremity with interesting non-robotic assisted therapy with Armeo® equipment.

Author details

Nina Sladekova[1], Elena Ziakova[1, 2]*, Jaroslav Kresanek[1], Stanislava Klobucka[2], Jana Havlova[1] and Miroslav Malay[3]

*Address all correspondence to: elenaziakov@gmail.com

1 Department of Physiotherapy, Faculty of Nursing and Professional Health Studies, Slovak Medical University, Bratislava, Slovak Republic

2 Rehabilitation Centre Harmony, Bratislava, Slovak Republic

3 Faculty of Nursing and Professional Health Studies, Slovak Medical University, Bratislava, Slovak Republic

References

[1] Riener R, Harders M. Chapter 7 – Virtual reality for rehabilitation. In Virtual Reality in Medicine. London: Springer-Verlag; 2012. p. 161–180. DOI: 10.1007/978-1-4471-4011-5

[2] Sivak S, Nosal V, Bittsansky M, Dluha J, Dobrota D, Kurca E. Type and occurrence of serious complications in patients after mild traumatic brain injury. Bratisl Med J. 2016; **117**: 22–25. DOI:10.4149/BLL-2016-005

[3] Dew AP, Moreau NG. Comparison of 2 techniques for measuring rectus femoris muscle thickness in cerebral palsy. Pediatr Phys Ther. 2012; **24**: 218–222. DOI: 10.1097/PEP.0b013e31825bcb77

[4] Gordon AM. What does selective motor control of the upper extremity in cerebral palsy tell us? Dev Med Child Neurol. 2016; **58**: 536–537. DOI: 10.1111/dmcn.13028

[5] Westcott MS, Yocum A, Bartlett DJ, Mendoza J, Jeffries L, Chiarello L, Palisano R. Development of the early activity scale for endurance for children with cerebral palsy. Pediatr Phys Ther. 2012; **24**: 232–240. DOI: 10.1097/PEP.0b013e31825c16f6

[6] Piña-Garza JE. Fenichel's Clinical Pediatric Neurology. 7th ed. United States of America: Elsevier Inc.; 2013. 412 p. ISBN: 978-1-4557-2376-8

[7] Hocoma. Armeo® User Manual [CD-ROM]. Switzerland: Hocoma AG; 2008.

[8] Berryman Reese N, Bandy WD. Joint Range of Motion and Muscle Length Testing. 2nd ed. United States of America: W.B. Saunders Company; 2009. 437 p. ISBN-13: 978–1416058847

[9] Kolar P, et al. Clinical Rehabilitation. 1st English ed. Praha: Alena Kobesova; 2013. 770 p. ISBN-13: 978–8090543805

[10] World Medical Association: Declaration of Helsinki – Ethical Principles for Medical Research Involving Human Subjects [Internet]. 1964 – rev. 2000. Available from: http://www.wma.net [Accessed: 2016-06-25].

[11] Ladenheim B, Altenburger P, Cardinal R, Monterroso L, Dierks T, Mast J, Krebs HI. The effect of random or sequential presentation of targets during robot-assisted therapy on children. NeuroRehabilitation. 2013; **33**: 25–31. DOI: 10.3233/NRE-130924

[12] Krebs H, et al. Motor learning characterizes habilitation of children with hemiplegic cerebral palsy. Neurorehabil Neural Repair. 2012; **26**: 855–860. DOI: 10.1177/1545968311433427

[13] Rudhe C, Albisser U, Starkey ML, Curt A, Bolliger M. Reliability of movement workspace measurements in a passive arm orthosis used in spinal cord injury rehabilitation. J Neuroeng Rehabil. 2012; **9**: 37. DOI: 10.1186/1743-0003-9-37

[14] Lo AC, Guarino PD, Richards LG, Haselkorn JK, Wittenberg GF, et al. Robot-assisted therapy for long-term upper-limb impairment after stroke. N Engl J Med. 2010; **362**: 1772–1783. DOI: 10.1056/NEJMoa0911341

[15] Klobucka, S., Kralovicova, M., Ziakova, E. A functionally robot-assisted therapy upper limb. Rehabilitation and Physical Medicine. 2010; **17**: 164–168.

[16] Bayon C, Raya R, Lerma Lara S, Ramírez O, et al. Robotic therapies for children with cerebral palsy: A systematic review. Transl Biomed. 2016; **7**: 44. DOI:10.21767/2172-0479.100044

[17] Grishkina MN, Guliaeva NV, Akzhigitov RG, Gersamia AG, et al. The effect of childhood stress on the development of mental disorders in adults. Journal of Neurology and Psychiatry S S Korsakova. 2015; **115**: 16–19.

[18] Müller C. Armeo Masterclass: Learn to Motivate Your Patients More than Ever Before! Hocoma [Internet]. 2015. Available from: https://www.hocoma.com/world/en/news-and-events/news article/article/armeo-masterclass-learn-to-motivate-your-patients-more-than-ever-before/ [Accessed: 2016-07-04].

[19] Li M, Lu S, Wang G, Feng L, Fu B, Emotion NZ. Working memory, and cognitive control in patients with first-onset and previously untreated minor depressive disorders. J Int Med Res. 2016; **44**: 529–541. DOI: 10.1177/0300060516639169

[20] Maciejasz P, Eschweiler J, Gerlach-Hahn K, Jansen-Troy A, Leonhardt S. A survey on robotic devices for upper limb rehabilitation. J NeuroEng Rehabil. 2014; **11:** 3. DOI: 10.1186/1743-0003-11-3

Chapter 7

After Stroke Movement Impairments: A Review of Current Technologies for Rehabilitation

Pablo Aqueveque, Paulina Ortega, Esteban Pino,
Francisco Saavedra, Enrique Germany and
Britam Gómez

Additional information is available at the end of the chapter

http://dx.doi.org/10.5772/67577

Abstract

This chapter presents a review of the rehabilitation technologies for people who have suffered a stroke, comparing and analyzing the impact that these technologies have on their recovery in the short and long term. The problematic is presented, and motor impairments for upper and lower limbs are characterized. The goal of this chapter is to show novel trends and research for the assistance and treatment of motor impairment caused by strokes.

Keywords: stroke, hemiparesis, rehabilitation, assessment technologies, upper limb, lower limb, FES

1. Introduction

Stroke is the most common acquired neurological disease in the adult population worldwide (15 million every year [1]). Based on recently published studies, incidence of stroke in Europe at the beginning of the twenty-first century ranged from 95 to 290/100,000 per year [37]. Between 2000 and 2010, the relative rate of stroke deaths dropped by 35.8% in the United States and other countries. However, each year stroke affects nearly 800,000 individuals, becoming the first cause of chronic disability and the third cause of death. It is a global public health problem worldwide that generates a significant burden of illness for healthy life years lost due to disability and premature death.

One-third of stroke survivors achieve only a poor functional outcome 5 years after the onset of stroke. Although there is great progress in the management of acute stroke, most of the care to reduce dependence on post-stroke patients depends on rehabilitation. Optimal functional recovery is the ultimate goal of neurorehabilitation after acute brain injury, mainly by optimizing sensorimotor performance in functional actions. New brain imaging techniques are making it clear that the neurological system is continually remodeling throughout life and after damage through experience and learning in response to activity and behavior.

Rehabilitation in stroke patients seeks to minimize the neurological deficit and its complications, encourage family, and facilitate social reintegration of the individual to ultimately improve their quality of life. Stroke rehabilitation is divided into three phases. The acute phase usually extends for the 1st weeks, where patients get treated and stabilized in a hospital and get stabilized. Subacute phase (1–6 months) is the phase where the rehabilitation process is more effective for recovering functions. In chronic phase (after 6 months), rehabilitation is meant to treat and decrease motor sequels.

The potential ability of the brain to readapt after injury is known as neuroplasticity, which is the basic mechanism underlying improvement in functional outcome after stroke. Therefore, one important goal of rehabilitation of stroke patients is the effective use of neuroplasticity for functional recovery [38].

As mentioned before, neural plasticity is the ability of nervous system to reorganize its structure, function, and connections in response to training. The type and extent of neural plasticity is task—specific, highly time-sensitive and strongly influenced by environmental factors as well as motivation and attention.

Current understanding of mechanisms underlying neural plasticity changes after stroke stems from experimental models as well as clinical studies and provides the foundation for evidence-based neurorehabilitation. Evidence accumulated during the past 2 decades together with recent advances in the field of stroke recovery clearly shows that the effects of neurorehabilitation can be enhanced by behavioral manipulations in combination with adjuvant therapies that stimulate the endogenous neural plasticity.

Nowadays, a large toolbox of training-oriented rehabilitation techniques has been developed, which allows the increase of independence and quality of life of the patients and their families [39]. The recovery of function has been shown to depend on the intensity of therapy, repetition of specified-skilled movements directed toward the motor deficits and rewarded with performance-dependent feedback.

The use of technological devices not only helps to increase these aspects but also facilitates the work of therapists in order to enhance the abilities of patients and a higher level of functional recovery. They create environments with a greater amount of sensorimotor stimuli that enhance the neuroplasticity of patients, translating into a successful functional recovery. The use of technological devices can transfer the effects of rehabilitation to the different environments where patients spend their daily life allowing a favorable social reintegration. In this chapter, a review of technologies for rehabilitation of mobility in upper and lower extremity is presented.

2. Motor impairments after stroke

One of the most important areas affected by stroke is motor skills. The patients may have disabilities in different degrees (mid, acute, severe), in different hemispheres (one or both), and at different levels: upper (face, neck), medium (trunk, upper limbs) and lower (lower limbs). Hemiparesis and motor recovery have been the most studied of all stroke impairments. Hemiparesis defined as muscular weakness or partial paralysis restricted to one side of the body is an impairment present in 88% of the stroke patients, affecting lower and upper limbs. Six months after stroke about 38% of patients lightly recovers dexterity in the arm and only 12% shows full recovery after conventional rehabilitation therapy [2]. Weakness and paresis are the most important impairments on the early stages after stroke as they lead to a learned nonuse of limbs. Immobility, chronic pain, and some sensory impairments can also contribute to the learned nonuse state. As the recovery progresses, spasticity and spastic co-contractions can induce some compensatory movements, which if are persistent in time and repeated may contribute to a learned bad use [3].

For healthcare organizations, it is difficult to assess in a general and accurate way the effects that a stroke can have on people given the variety of areas involved. However, there are tools that assessing in a global way the degree of disability of the condition, regardless of the area where the impairment is found. In recent times, there are new technological tools to obtain data and also there are several tests that evaluate upper limb motor functions, trunk functions, gait capacity, and spasticity. This allows health professional determine diagnosis and appropriate therapeutic interventions.

3. Upper limb problems

Many of the activities we do during the day involve using the upper limbs such as when eating, dressing, and writing. Their use is not only associated with the use of everyday instruments but also with the contact with the world and the way we interact with other people. The accomplishment of these tasks requires sequences of complex movements that integrate the activation of appropriate muscular groups and the sensorimotor coordination of the hands, which translates into an effective functional action.

Grasp and manipulation are strategies of movement that are mainly affected in stroke patients. Recent studies have found that recovery is minimal in some individuals, particularly those with a flaccid paretic limb in the first few weeks. This is why dysfunction in upper limbs is a major clinical, economic, and social problem for neurorehabilitation teams. Hemiparesis on upper limb usually affects the hand causing weakness and spasticity, leading to a decrease in movement precision, muscle fatigue, lack of coordination, and an impaired ability to grasp objects, having a great impact on daily living activities [41].

Impairments such as a decreased motor impulse, a lower frequency of neuronal activation, poor sequencing and coordination of segmental movements, and sensory deficits have a marked influence on the functional performance of the upper limb. Muscle weakness and loss of manual dexterity may be accompanied by the development of soft tissue changes and shoulder pain.

Many studies have shown that increasing therapy time in the upper limb from the acute stage reduces associated impairments and improves function satisfactorily from a clinical standpoint. This must be related to an intensity and dose of therapy appropriate to generate substantial changes.

It have been shown that patients have a better motor function when performing a specific task involving a useful interaction with an object, practice of strengthening exercises and functional actions is as important after stroke as for anyone attempting to gain strength and ability in motor actions [40].

3.1. Treatment-oriented devices and assistive devices for upper limb rehabilitation

In upper extremity rehabilitation, we can make a distinction between two categories of technologies. In the first category, we have treatment-oriented devices that are used to assist the exercises during the early rehabilitation process. The second category is made up of assistive devices, designed to aid patients in their daily living activities. Both categories may apply different therapeutic approaches, such as constraint-induced therapy, biofeedback therapy, and robot-aided therapy.

3.1.1. Treatment-oriented devices

This type of devices aims to help therapists in the flaccidity stage of hemiparesis due to the lack of neurologic connections. There is an agreement between therapists that early focused and repetitive exercise is the most important aspect for future recovery [4]. There is also evidence that early rehabilitation treatments may induce muscular reinnervation processes that can recover motor functions [5]. This category device can be subdivided into two more groups. The first consists mainly on mechanical structures to bring support to the limb and set constraints for the movements, like The Armeo® by Hocoma, and the Saebo ReJoyce (see **Figure 1**). They are used for exercising purposes promoting in this way the reinnervation of muscles in the affected limb.

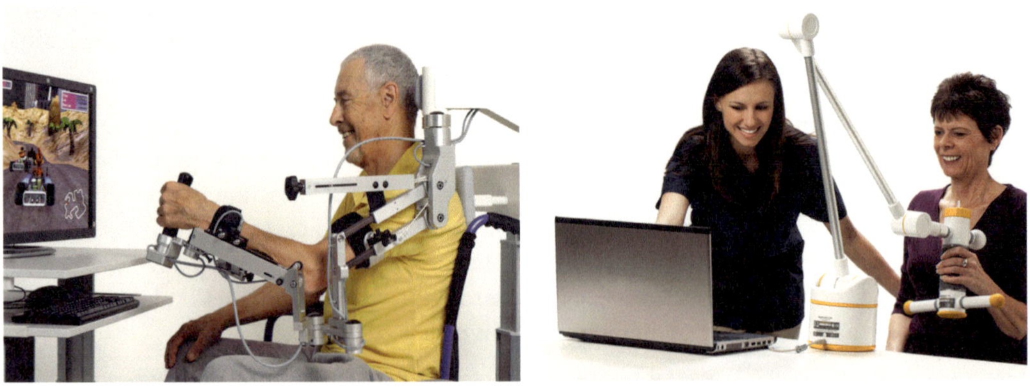

a) b)

Figure 1. Mechanical treatment devices. (a) Armeo Spring and (b) Saebo ReJoyce.

Armeo Spring is an arm and hand rehabilitation exoskeleton with 5 degree-of-freedom (three in the shoulder, one in the elbow, one in the forearm) orthosis, which covers an individual's arm to protect, achieve a larger active range of motion and automatically guide his/her therapy by applying repetitive movements in an environment in which the subject is stimulated by interactive rehabilitation methods (videos, games, and instructions). The goal of this system is to restore the movement and functionality of the affected limb in less time than conventional therapeutic methods. This system is used in individuals who have suffered from strokes, traumatic brain injury, or other neurological disorders that induce hand and arm impairments [6].

Saebo Rejoyce is an upper limb training system used in people who have suffered upper limb impairments after strokes or other neurological conditions. It is a gaming system to practice gross and fine motor tasks, simulating functions like opening doors, opening jars, grasping and turning a cup, turning a key, and manipulating coins and other small objects, assessing and tracking his performance in order to generate a progress report by session [7].

In the second group, we find therapeutic stimulators like MyoTrac Saebo Infiniti and Neuromove 900, which use functional electrical stimulation (FES) to exercise the grasping movements based on the acquisition of electromyography signals to trigger the stimulation pulse, in order to enable a more natural control using the user movement intent (see **Figure 2**) [8].

MyoTrack Infinity is a device used to reach the skeletal muscle re-education and the rehabilitation of the arm, or other parts of the body, after strokes (or other affection), through the application of electrical stimulation by the measure of a high-resolution surface electromyography.

Neuromove 900 is an electrical stimulator triggered by electromyography, sensing the muscular activity through reusable surface electrodes. The device evaluates the activity present in the muscle and then sets a higher standard that the patient should try to reach. Upon reaching the threshold, the patient is rewarded with electrical stimulation that makes the muscle move for a few seconds. Success is measured in the actual movement and gives the patient greater control over his/her extremity. It is used for stroke rehabilitation, spinal cord injury, manual, or stim only.

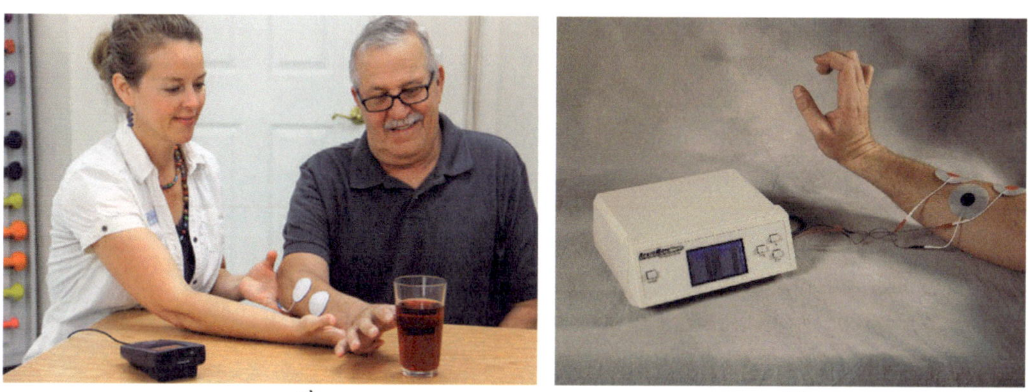

a) b)

Figure 2. Treatment devices with FES capabilities. (a) Saebo MyoTrack Infinity and (b) Neuromove 900.

3.1.2. Assistive devices

These types of systems are designed to aid sequel patients in their daily life activities, due to chronic conditions and lack of normal movement and functions. Besides the assistive goal of these devices, they also seek to promote a long-term recovery process. As expected these types of devices are designed to be portable and wearable. We could also subdivide this category into two: the one that uses stimulation and the mechanical only device. Some commercially available devices are the Bioness NESS H200 and the Saebo Glove. The Bioness NESS H200 is an FES device that consists of an orthosis that stabilizes the wrist of the affected limb to stimulate and contract the hand. Once positioned, this device allows grasp, hold, and release by neurostimulation. The Saebo Glove and Aider Stroke Rehabilitation Glove (see **Figure 3**) are glove-shaped devices that allow the extension of the hand by the action of elastic materials, generating support in patients who present a decreased grasping strength. These systems are used in patients that can only generate a limited closing grasping function and have almost fully compromised its ability to extend-back the fingers.

Bioness NESS H200 is an electronic device that consists in an orthosis and his control unit. The control unit transmits electrical pulses to the peripheral nerves through electrodes into the orthosis, activating five muscle groups of the forearm and the hand to generate the grasping moves of the hand. The advantage of this system is that it provides a proprioceptive input to the user, facilitating the normal control of the movements of the hand. Other benefits of the use of this device are reducing the muscle spasms, increasing or maintaining range of the movements, improving the blood circulation, and retarding the atrophy of the muscle.

SaeboGlove is an orthopedic glove that facilitates the movements of grasping of the hand supporting the extension of the fingers by elastics whose tension is adjusted according to the characteristics of the individual. This system is fixed to the hand by means of two straps placed in the hand and in the forearm. It can also be combined with electrical stimulation techniques to support the hand closing movement. Individuals who had stroke, brain injury, brachial plexus injury, radial nerve palsy, individuals with limited wrist and finger extension are eligible to use this system.

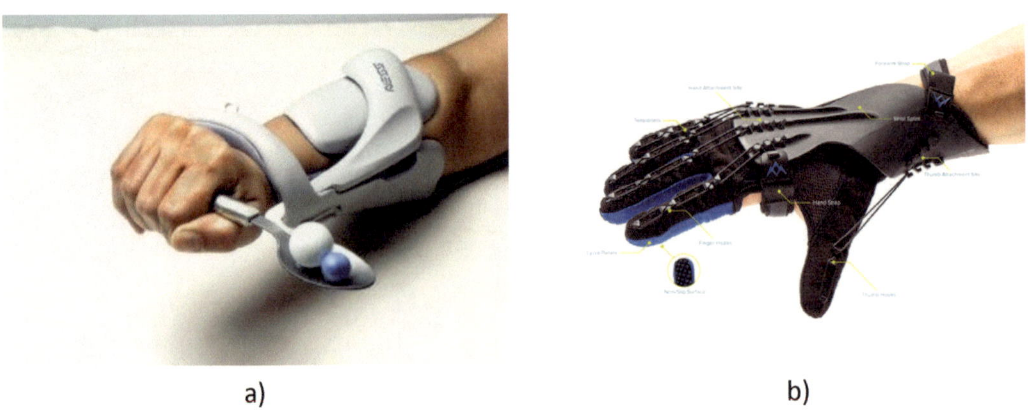

a) b)

Figure 3. Assistive devices for daily living support. (a) Bioness NESS H200 and (b) Saebo Glove.

3.2. Upper limb assessment indexes and tests

During assessment of patients, there are different indexes that therapists use to evaluate the capabilities of the upper extremity functions. Reviewing literature and clinical trials, the most used are as follows:

The *Fugl-Meyer assessment (FMA)* is a stroke-specific, performance-based impairment index. It is designed to assess motor function, balance, sensation, and joint function in patients with post-stroke hemiplegia. It is applied clinically and in research to determine disease severity, describe motor recovery, and to plan and assess treatment.

The *action research arm test* (MAS) is a performance-based scale to assess everyday motor function in patients with stroke. The MAS is based on a task-oriented approach, assessing performance of functional tasks rather than isolated patterns of movement.

The *Chedoke-McMaster stroke assessment* measures physical impairment and disability in patients with stroke and other neurological impairment. The measure consists of an impairment inventory and an activity inventory. The first inventory aims to determine the presence and severity of common physical impairments, to classify or stratify patients when planning, selecting interventions, and evaluating their effectiveness and to predict outcomes. The second inventory measures changes in physical function. The Chedoke-McMaster stroke assessment is a discriminative, predictive and evaluative tool.

The *box and blocks test* (*BBT*) is a functional test used in upper limb rehabilitation. The test is used to measure the gross manual dexterity of a patient or of a person using an upper limb prosthetic device.

3.3. Clinical evidence supporting technological devices in upper limb rehabilitation

Many studies regarding these types of systems have been reported in the literature, they all base their finding on randomized controlled trials, involving stroke patients to short-term therapies using this kind of devices and a control group subject to conventional therapy to determine whether rehabilitation protocols that use these technologies are improving the functional aspects and quality of life of stroke survivors with upper-limb deficits. In **Table 1**, we summarize some recent studies that compare hand function improvements between groups in rehabilitation with and without technological support.

As we can observe from the results above, the groups that were treated with this type of devices showed greater improvements in some of the evaluated aspects than the ones that received traditional manual therapy. Devices that offered mechanical support and mechanical movement assistance did report an increase in strength and hand reaching. This improvement can be explained by muscle recovery from atrophy state due to cyclical exercising and also to the reduced force needed to maintain the arm position from the gravity compensation, effort that can be focused on extension muscles. Other important factor to notice is the usage of FES to aid the movements in some devices. Almost every device with FES capabilities reported an improvement compared to the control groups in hand opening and reaching. An important factor to notice here is that the benefits from FES were maintained after the

Reference	Hemiparetic side	Post-stroke time	N of patients	Intervention time	Device type	Assistance type	Improvements
Krabben et al. [9]	Single side	12 months (AVG)	8	–	Assistive w/FES	GC* and FES	Hand opening
Kim et al. [10]	Single side	<6 months	30	5 weeks	Treatment w/FES	EMG-triggered NMES	RT** of hemiparetic wrist
Knutson et al. [11]	Single side	<6 months	6	6 weeks	Treatment w/FES	CCFES***	Dexterity, MAFEA****, reaching, hand opening
Knutson et al. [11]	Single side	>6 months	21	12 weeks	Treatment w/FES	CCFES***	MAFEA****
Fischer et al. [12]	Single side	<6 months	15	5 weeks	Treatment	Cyclic stretching	Strength gain,
Makowski et al. [13]	Single side	>6 months	3	–	Treatment w/FES	Arm support and FES	Hand opening and reaching

* GC, gravity compensation.

** RT, reaction time.

*** CCFES, contralaterally controlled functional electrical stimulation.

**** MAFEA, maximum active finger extension angle.

Table 1. Clinical reports for upper extremity improvements compared with traditional therapy.

interventions (follow-up tests) only when the stimulation was triggered by the user itself, such as EMG-triggered or contralaterally controlled therapy. The main idea that resides here is the treatments that involve cognitive awareness from the user generate favorable conditions in rehabilitation and promote faster motor control recovery. Motivation is a key factor that determines the achievements during rehabilitation therapies; in this way, devices that maximize cognitive engagement during the process are the ones that report the most improvements in contrast with traditional therapy. EMG-triggered devices with FES capabilities are the most promising regarding clinical results. Even though many of the clinical reports with use of technological assistive devices are relatively short-term evaluations, trends in rehabilitation achievements are positive in almost every aspect of the assessment indexes used. Long-term evaluation and clinical reports are underway as stated in many of the cited publications, considering also a most representative number of patients to obtain statistical data that would allow to validate the favorable trends observed in these studies.

4. Lower limb problems

The ability to walk independently is a prerequisite for most activities of daily living. The ability to walk in a community environment requires the ability to walk at speeds that allow the individual to cross the street in the time set by traffic lights, under or over objects or handling curbs.

http://dx.doi.org/10.5772/67577

Gait dysfunction is common in individuals with neurological disorders not only due to injury-related disorders but also to the cardiovascular and musculoskeletal consequences of disuse and physical inactivity. Impairments following stroke usually involve an excessive energy cost during walking, which limits the type and duration of activities. Stroke patients are generally unable to comfortably maintain the most efficient walking speed beyond a very short distance [42]. These individuals are often restricted only to activities of daily living within confined spaces in the home and with little possibility of performing functional activities at community level, which restricts their social participation and ultimately affects their quality of life and independence. Some of the limitations that can be observed in patients with stroke are related to poor motor control, muscle weakness and/or soft tissue shortening, sensory and balance disturbances, among others. These impediments are specifically translated into pathological characteristics of gait that affect the proper sequential activation of the muscles in the different stages of gait, causing compensatory strategies that decrease gait speed and efficacy and that may increase the risk of falls from the patients.

One of the most common impairments observed in walking of stroke patients is the reduction of ankle dorsiflexion during hill contact and during the support phase associated with a hyperextension of the knee, which results in a fall of the foot during gait. This may be due to decreased activation of the anterior tibial muscle, as well as to premature activation of the calf muscles. This condition not only affects the gait speed but also limits the ability to walk in irregular ground and surfaces and go up and down stairs. This impairment is commonly known as drop foot. Drop foot is a disorder characterized by a lack of voluntary control of dorsiflexor muscles. A person with foot drop cannot lift the front of the foot before the heel comes in contact with the ground, which can cause tripping or falling. As a result, patients develop compensatory strategies including pelvic obliquity, hip hiking, and hip abduction with circumduction gait pattern to preserve foot clearance, see **Figure 4**. These strategies increase significantly the energy consumption of the person. With the objective of improving gait efficiency and safety, and overall improvement of the gait pattern to reduce musculoskeletal stress from altered biomechanics, many treatment modalities have been used. Treatment modalities include stretching, exercise, rehabilitation, orthotics, and assistive devices.

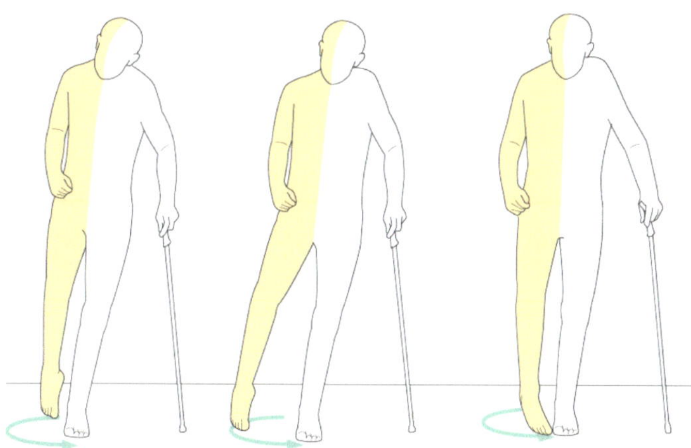

Figure 4. Walking pattern in person with drop foot [14].

4.1. Assistive technologies devices for lower limb rehabilitation

4.1.1. Ankle foot orthosis (AFO)

Regarding assistive-type technologies, the common method for the treatment of the droop is ankle-foot orthoses (AFO). In general, AFOs stabilize the foot and ankle by lifting the tip of the toes when the foot loses contact with the surface while walking, providing stability, control, and protection for the foot. Moreover, this assistive tool grants proprioceptive feedback to know the position that the foot has in space.

There are a large number of AFOs, which differ from one other according to the medical and biomechanical needs of the individual. Some types of AFOs are flexible ankle-foot orthoses, hinged ankle-foot orthoses, tubular ankle-foot orthoses, silicone ankle-foot orthoses (SAFO), Charcot restraint orthotic walker, and plantar fasciitis night-splint (see **Figure 5**).

Flexible ankle-foot orthoses are lightweight orthoses whose design prevents the foot from performing plantar flexion, preventing its fall, allowing a smooth sway of the foot without the finger dragging on the floor. In general, the shape of these orthoses depends on the needs of the user. Flexible AFOs are made with propylene plastic, with a Velcro that allows closing and fixing the orthosis to the leg. They are used for people who have suffered a stroke, multiple sclerosis, poliomyelitis, or other nerve damage.

Hinged AFOs are effective elements that are used for the control of plantar flexion, dorsi flexion, and lateral movements. Many designs are used with a wide variety of hinges, which are selected based on the requirements of the user and considerations regarding the weight and shape of the device. Hinged AFOs are used for people who suffer from droop foot such as strokes (CVA) or cerebral palsy. They are generally made of plastic.

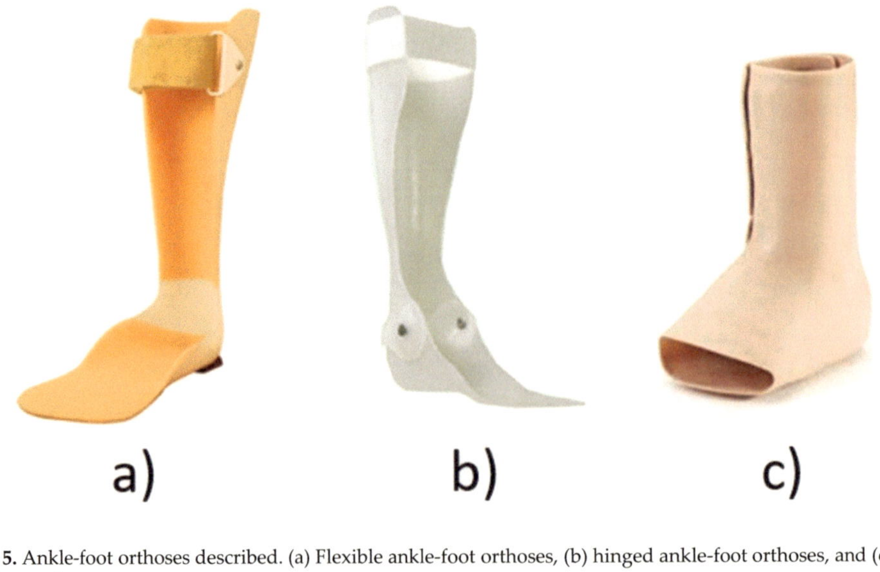

Figure 5. Ankle-foot orthoses described. (a) Flexible ankle-foot orthoses, (b) hinged ankle-foot orthoses, and (c) silicone ankle-foot orthoses.

http://dx.doi.org/10.5772/67577

Tubular AFOs also called circumferential AFOs enclose the leg and foot completely. The detachable strap provides great stability and protection for the leg and foot. The clamp is lined with foam and antimycotic leather to add protection for sensitive skins. Tubular AFO is most commonly used for people with diabetic complications or other peripheral neuropathies.

SAFO is a more current design, used for people who have flaccid foot paralysis as a sequel to pathologies such as Charcot-Marie-tooth disease, multiple sclerosis, poliomyelitis, stroke, and spinal cord injury. Its design offers an optimal proprioception so that the user can know and feel in a more comfortable way the position of his foot in space, giving a great control to the plantar flexion. The orthosis has a very low profile, adapting to the shape of the individual's foot, allowing it to be worn with or without shoes.

These types of elements have proven to be very useful, significantly improving the dynamics with which the gait exercise is performed [15–17]. Studies show that regardless of the material and the type of AFO used, the measurable parameters for gait exercise (cadence, joint angles, balance) do not differ from each other, the most important being the comfort and security that these elements confer to the individual at the time of choosing them.

Lately, control systems have been developed for AFOs in order to dampen the force with the foot reaches the ground in the support stage of the gait and generates a certain level of dynamics in the movement of the foot adjusting the rigidity between the foot and ankle. These systems are called *active ankle-foot orthoses* (*AAFO*) (see **Figure 6**). The control system, developed by Joaquin A. Blaya and Hugh Herr, works by applying a dynamically controlled torque to the orthosis joint, to cushion the shock of the foot on the floor while

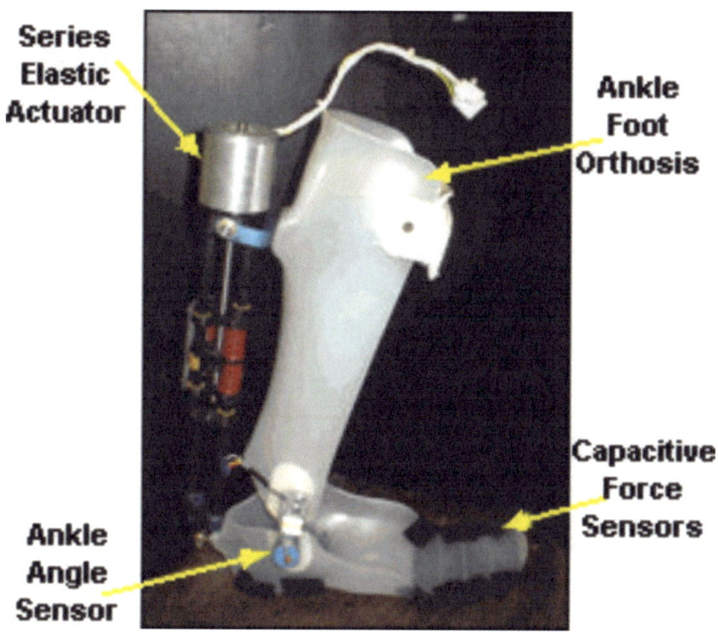

Figure 6. Active ankle-foot orthoses (AAFO) developed by Joaquin A. Blaya and Hugh Herr [18].

the foot is falling. Then, after the placement of the foot, the control system minimizes the stiffness of the orthosis to allow plantar flexion of the subject until the foot loses contact with the ground entering the swing phase of the gait, where a constant stiffness is applied to force the dorsiflexion of the foot. Tests show that the active stiffness adjustment reduces the occurrence of slap foot, allows greater powered plantar flexion, and provides for less kinematic difference during swing compared to normal, presenting a system that has clinical benefits compared to the conventional orthoses [18].

However, AFOs have a number of disadvantages and limitations, since these devices do not promote active movement, can be uncomfortable, bulky and, if misplaced, produce areas of pressure pain and tissue breakdown. This is why functional electrical stimulation (FES) is used as an alternative to AFOs.

4.1.2. Functional electrical stimulation (FES)

Functional electrical stimulation is a rehabilitation technology that uses electrical periodic pulses to stimulate the nerves that produce contractions in paralyzed muscles and recover lost functions. There are two ways to apply FES: surface FES and implanted FES.

4.1.2.1. Surface FES

Liberson in 1961 [19] proposed the use of FES to correct drop foot. He created a system that uses electrical stimulation applied to the common peroneal nerve, recruiting muscles involved in dorsiflexion and eversion of the ankle. The operation of the system is as follows: two superficial electrodes are located just below the head of the fibula bone. Using a foot switch placed in the shoe under the heel, the stimulation pulses are synchronized with the gait cycle. Then, the foot is lifted through the swing phase. The typical stimulation profile (see **Figure 7**) is often a ramping up and down of the stimulus. The rise and fall of the stimulation envelope can be adjusted to prevent a stretch reflex in the calf muscles and to prevent foot flap due to the premature ending of dorsiflexion.

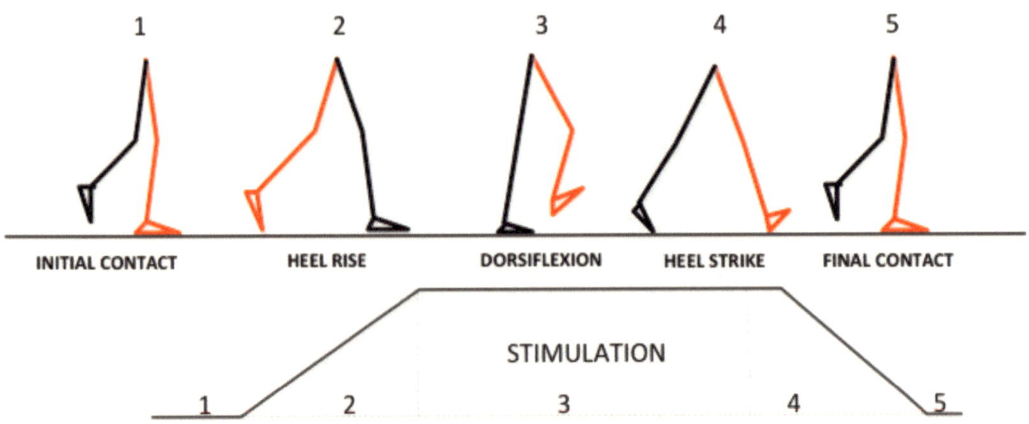

Figure 7. Typical stimulation pulse profile on drop foot FES in each stage of the gait cycle.

The use of these devices can be classified in two ways [20]: the first is *the orthotic effect* that is the direct effect of using the FES (lift the toe). The second effect is the therapeutic effect and relates to changes in walking ability when not using FES. Although not widely used or universally available, there is growing evidence that treatment with FES improvements in gait speed, cadence, improved confidence in walking, reduction in the risk of falling, less effort during walking, and the active contraction produced by FES can help to prevent muscle atrophy.

Since, many research groups have designed and studied new stimulators that are based on the work of Liberson, creating FES devices currently commercially available, such as the Odstock dropped foot stimulator (ODFS®, Odstock Medical Limited, Salisbury, UK), the WalkAide® system (Innovative Neurotronics Inc., Austin, TX, US), the Bioness NESS L300® foot drop system (Bioness Inc., Valencia, CA, US), and the MyGait® system (Ottobock, Duderstadt, Germany).

Odstock dropped foot stimulator: The ODFS is single channel; it is controlled with a wired foot switch placed in the shoe, which is used to turn on and off the stimulation at the right time while walking (see **Figure 8a**). In 2016, a wireless version of the footswitch began to be commercialized. Typically, skin surface electrodes are placed over the common peroneal nerve as it passes over the head of the fibula bone and the motor point of the tibialis anterior.

Walkaide system: The Walkaide is also cuff-based surface stimulator (see **Figure 8b**). The main difference from other systems is that it uses an inertial gait sensor to trigger the stimulation based on acceleration thresholds. However, it has the option of a wired foot switch. Internally, it has an algorithm that auto-adjusts the parameters for each patient.

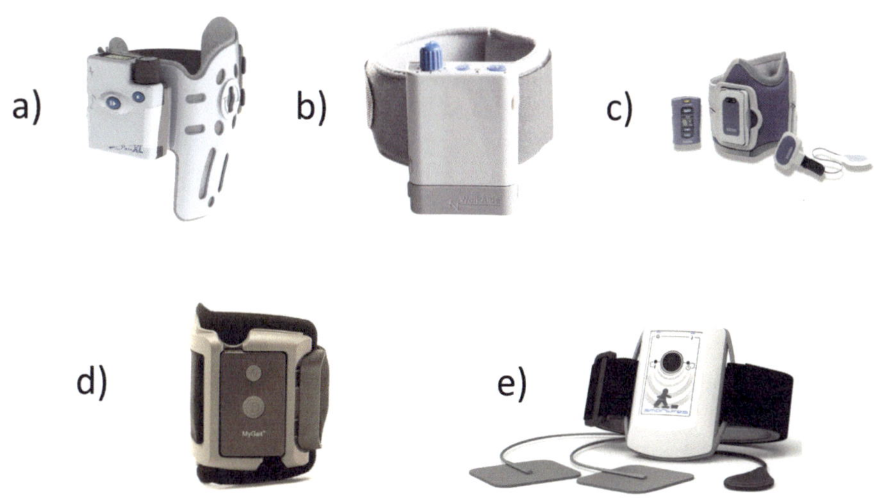

Figure 8. Surface FES systems available commercially: (a) ODFS system [21], (b) Walkaide System [22], (c) My Gait System [23], (d) Ness L300 [24], (e) SmartFES System [43].

Ness L300 system: The Ness L300 is a surface FES system that includes the stimulation unit based in a cuff, a wireless foot switch, and hand-held control for intensity control (see **Figure 8c**). Recently, the L300 plus was released. This consists of a thigh cuff to give you greater control over bending and straightening your knee, which may help you walk more naturally.

My Gait: Is the newest surface stimulator launched in 2013. It is a cuff-based surface stimulator (see **Figure 8d**). The novelty of this stimulator is that it provides two channels. The second channel is integrated and does not need an additional device. By stimulating additional muscle groups, gait performance can be further improved. It can be employed to support in flexion or extension of the knee, improved triggering of the swing phase, and minimized compensatory movements.

SmartFES: Surface stimulator launched in 2016. The SmartFES Is a low cost single channel stimulator, it is controlled with a wired foot switch placed in the shoe. The main difference from other systems is the Unit Interface (UI). The UI is implemented in an Android application and allows to modify the stimulation parameters in order to obtain the best response of dorsiflexion. The unit interface communicates via Bluetooth with the stimulator. This allows an easy and fast operation. Actually the SmartFES is being commercialized in Chile.

Table 2 summarizes the technical characteristics of the systems mentioned above.

Many research studies have demonstrated that FES combined with conventional physio leads to further improved results, compared to conventional therapy alone [25]. Taylor et al. test the ODFS stimulator in 32 subjects [26]. The results show that 71% of ODFS users were able to walk further. Additionally, it was reported that 33% of users used the device to keep them fit, 70% used the

	Ness L300®	**Walkaide®**	**ODFS Pace®**	**MyGait®**	
Manufacturer	**Bioness Neuromodulation Ltd.**	**Innovative Neurotronics Inc.**	**Odstock Medical Ltd**	**Ottobock Healthcare Products GmbH**	**Neurotech Spa. SmartFES**
Amplitude (mA)	0–80 @ 1 k	0–121 @ 1 k	10–100 @ 1k	0–90 @ 1k	10–100 @ 1kΩ
Number of channels	1	1	1	2	1
Frequency (Hz)	20–45	16.7–33	0–60	10–80	10–60
Pulse width (us)	100/200/300	25–300	0–360	50–400	10–300
Foot switch	Wireless	Inertial/wired	Wired/wireless	Wireless	Wired Foot Switch
Waveform	Biphasic symmetrical/ asymmetrical	Biphasic asymmetrical	Biphasic symmetrical/ asymmetrical	Biphasic symmetrical/ asymmetrical	Biphasic Asymmetrical
Regulated current or regulated voltage	Regulated current	Constant voltage	Regulated current	Regulated current	Regulated Current
Battery	1 built in	1 replaceable	1 replaceable	1 built in	1 built in
Rechargeable	Yes	No	No	Yes	Yes
Country	USA	Canada	UK	Germany	Chile

Table 2. Commercial surface FES stimulator to correct drop foot.

ODFS for shopping and trips out, 57% used for social events, 19% for work; 79% users reported that their confidence was increased when walking and 52% reported that their independence was increased. Van Swigchem et al. [27] reported similar increases in gait velocity and cadence of 26 stroke patients with the Ness L300. Stein et al. [28] did an analysis with the Walkaide system of the orthotic and therapeutical effect among patients with a nonprogressive (stroke) and progressive (multiples sclerosis) disease. The mean walking speed at all times was typically 5–15% higher with the device than without the device for both the progressive and nonprogressive groups. Therefore, the immediate improvement is referred to as an orthotic effect. With respect to long-term changes as therapeutic effects, after 11 months the therapeutic effect was much larger for the nonprogressive group (about 30%) than for the progressive group (about 5%). It is necessary to make clear that the use of FES is not appropriate for all stroke patients. The patient has to be well motivated, able to walk with assistance or alone, and the muscle that raises the foot must not be denervated.

Despite the benefits of these devices, the surface FES devices have some disadvantages. When using surface electrodes, the lack of selectivity of muscles and nerves affected by superficial electrodes is a problem. The stimulation could be felt as "pins and needles" that in some cases can cause pain or discomfort. The other problem reported is skin irritation [29], and these cases are usually treated by changing the type of electrodes or modifying the stimulation settings. These problems can be solved by using implanted stimulators.

4.1.2.2. Implanted FES

Implanted devices eliminate the need to position electrodes on the skin each day and reduce all problems associated with surface stimulation. No soft tissue or skin reactions, no need for technically challenging electrode placement and no discomfort or pain due to constant electrical sensation through the skin.

These stimulators work by activating directly the nerve that controls the lifting of the foot, called common peroneal nerve. At a point, just below the knee, this nerve splits into two branches: the deep branch and the superficial branch. The deep branch goes to the muscles that lift (dorsiflex) and turn inward (inversion) the foot, while the superficial branch innerves the muscles that turn the foot outwards (eversion). In normal walking, a combination of these movements is required. Like a surface stimulator, the implanted stimulator uses a foot switch to detect the step and an external FES device activates the implant through a wireless antenna worn on the outside of the body.

Currently, there are two implantable devices to correct drop foot available in the market: the STIMuSTEP system (Finetech Medical Ltd., Welwyn Garden City, UK) and the ActiGait system developed by Neurodan A/S (Aalborg, Denmark), a subsidiary of Ottobock group (Berlin, Germany).

STIMuSTEP system: This system is a passive implantable dual channel peroneal nerve stimulator triggered by a foot switch (see **Figure 9a**). Communication between the external control unit and the foot switch is wired. Electrodes are surgically inserted, one in the deep branch and other in the superficial branch enabling the movements (dorsiflexion, eversion end inversion) to be controlled separately.

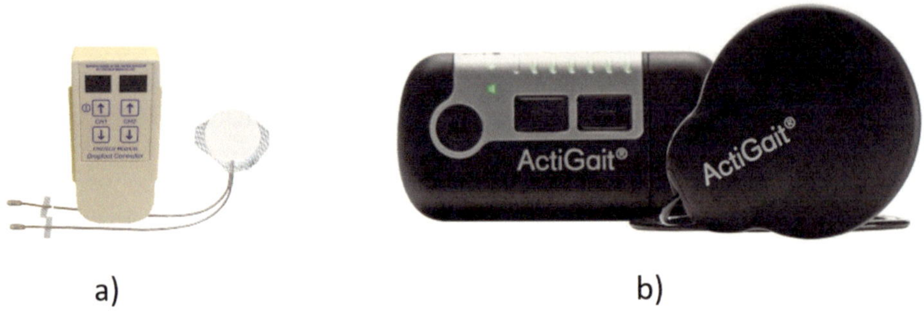

Figure 9. Implanted FES systems available commercially: (a) STIMuSTEP System [31] and (b) ActiGait System [32].

ActiGait system is an implantable four-channel nerve stimulator with a 12-contact electrode cuff (see **Figure 9b**). It works with a foot switch, which is worn in a sock, which triggers the initiation and termination of each stimulation sequence by a radio frequency wireless signal to the external control unit. The nerve stimulator contains a receiver for power and control and transmits the stimulation to the 12-electrode cuff through a subcutaneous cable [30]. The four channels can be programmed to allow for selective nerve bundle stimulation and balanced dorsiflexion/eversion.

Table 3 summarizes the technical characteristics of the systems mentioned above.

In general, surface and implanted FES reported similar results. The prescription of an implantable peroneal nerve stimulator is only a treatment option when the main goal is to achieve an orthotic effect for the long term in drop foot patients [32]. Burridge et al. [33] reported significant increases in mean gait distance and velocity (19%) using the ActiGait system. In a research conducted by Taylor et al. [26] 46 drop foot patients were selected from ODFS system users that had skin irritation, difficulties with electrode placement or anticipated long-term use with the device. Increases of 18% in gait velocities were reported. Additionally, patients also reported a more comfortable gait. In general, these devices are safe; however, an implant always has the risk of infection/rejection.

	ActiGait	**STIMuSTEP**
Manufacturer	Developed by Neurodan A/S a subsidiary of Ottobock Medical Ltd.	Finetech Medical Ltd.
Amplitude (mA)	1.2	16
Number of channels	4	2
Frequency (Hz)	5–50	30
Pulse width (us)	0–300	300
Sensor	Wireless foot switch	Wired foot switch
Waveform	Biphasic symmetrical	Biphasic asymmetrical
Regulated current or regulated voltage	Regulated current	Constant voltage
Country	Denmark/Germany	UK

Table 3. Commercial implanted FES stimulator to correct drop foot.

4.2. Robotic devices

Another technology that improves gait rehabilitation is robotic devices. These devices are characterized by providing safe, intensive and task-oriented rehabilitation. Lokomat Hokoma and the G-EO system (Reha-Technologies, Germany, GT) are some examples of low-limb robotic rehabilitation devices, which base their operation on restoring movement through repetitions in an assisted way, improving muscle strength, coordination, and locomotor retraining, reducing the rehabilitation, and/or recovery times of patients with stroke compared to methods based only on conventional kinesiological treatment techniques [34].

Lokomat is an exoskeleton consisting of a treadmill, a harness that allows different degrees of support of body weight and articulated arms that are placed embracing both legs (see **Figure 10**). These electromechanical arms mobilize hips and knees to perform the proper walking movements on the treadmill, in which the patient actively intervenes according to his possibilities [35]. It offers a physiological gait pattern with constant feedback and therapy assessment. It improves patient outcomes by increasing therapy volume and intensity, providing task-specific training, and increasing patient engagement.

G-EO SYSTEM is a device that consists of a harness, pedals for the user's feet, a central acquisition and control, where you can select the threshold levels for rehabilitation and incorporate two modes (see **Figure 11**). This is one of the most modern systems and has the capacity to simulate a walking on a level and on a ladder. It has two modes of operation. The first one, active mode, helps the patient to self-initiate the gait activity when a pre-selected mechanical

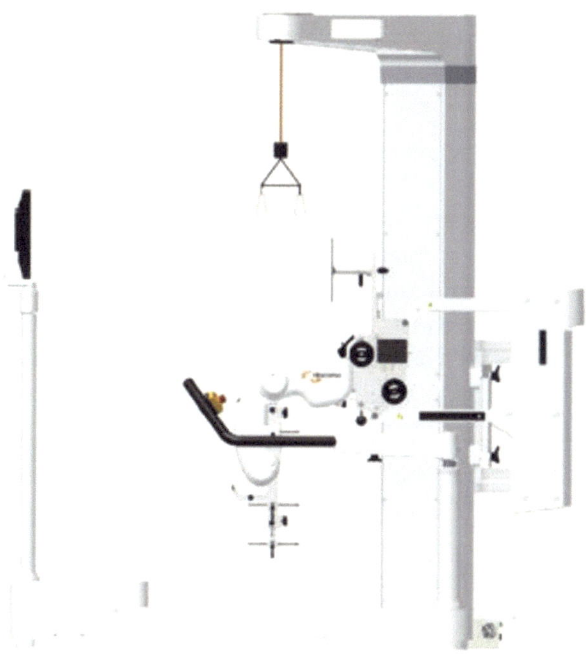

Figure 10. Lokomat®Nanos, using the same principles as LokomatPro Hocoma, but more compact.

Figure 11. The G-EO system for gait rehabilitation.

resistance "threshold" was adjusted. The second one, active-assistive mode, senses the patient's efforts to overcome the preselected-resistance threshold and then augments the patient's effort during the initiation of their gait movement [36].

Both systems previously described allow patients to regain their ability to walk in less time when compared to conventional therapy approaches, but it is strongly linked to the follow-up and type of therapy to which they are subject, since there is also evidence that indicates poor results [34].

5. Conclusions

The stroke (ACV) is a condition that in most cases leaves people who suffered it in an invalidating condition, ranging from cognitive problems to physical problems. From the point of view of physical affections, a great number of technologies and devices have been implemented that support not only the recovery and treatment of people but also their daily chores. In this chapter, a large number of assistive and rehabilitation systems have been presented, from the simplest ones like gloves, that serve to assist the opening of the hand, to more advanced systems like exoskeletons that are focused on facilitating complex movements in large degrees of freedom through interactive rehabilitation techniques.

With respect to the assistive systems, there are focused on recovering a certainly lost function of the user, allowing the increase of its independence. At present, a large number of systems based on functional electrical stimulation have been developed, applying electrical pulses to the muscles and/or nerves that are involved in the control of the intended movement. Recent applications seek to make these systems more portable, even moving to implantable solutions.

For the rehabilitation devices, it can be seen that the developing systems focused on rehabilitation are aimed at the creation and implementation of automated care methods that allow keeping the user motivated and thus improve their time of rehabilitation and/or recovery.

http://dx.doi.org/10.5772/67577

However these systems are promising, their performance will be strongly linked to the program that the professional in charge applies with them.

Acknowledgement

This work was supported by the National Fund for Scientific and Technological Development, FONDECYT, Chile, through FONDECYT N° 1140469

Author details

Pablo Aqueveque*, Paulina Ortega, Esteban Pino, Francisco Saavedra, Enrique Germany and Britam Gómez

*Address all correspondence to: pablo.aqueveque@udec.cl

University of Concepción, Concepcion, Chile

References

[1] WHO. World Health Organization, 2004. The Atlas of Heart Disease and Stroke, Part Three: Global Burden of Stroke.

[2] Twitchel, T.E. The restoration of motor function following hemiplegia in man. Brain, December 1951, vol. 74, no. 4, pp. 443–80. DOI: 10.1093/brain/74.4.443.

[3] Heart & Stroke Foundation, Functional Electrical Stimulation, Upper Extremity. Stroke Engine. Available from: http://www.strokengine.ca/patient-info/functional-electrical-stimulation-upper-extremity-info/. [Accessed: 2016-12-22].

[4] Post-Stroke Rehabilitation, National Institute of Neurological Disorders and Strokes, National Institute of Health, 2016. Available from: https://stroke.nih.gov/materials/rehabilitation.htm. [Accessed: 22-Sep-2016].

[5] Dobkin, B.H, Dorsch, A. New evidence for therapies in stroke rehabilitation. Current Atherosclerosis Reports, 2013, vol. 15, no. 6, pp. 1–9.

[6] Gijbels, D., et al. The Armeo Spring as training tool to improve upper limb functionality in multiple sclerosis: a pilot study. Journal of Neuroengineering and Rehabilitation, 2011, vol. 8, no 1, p. 1.

[7] Amanda, S. Product Review – Saebo Rejoyce. Posted 24 February 2015. Available from: http://www.archealth.com.au/post/2556994-product-review-saebo-rejoyce [Accessed: 2016-12-15].

[8] Hara, Y. Neurorehabilitation with new functional electrical stimulation for hemiparetic upper extremity in stroke patients. Journal of Nippon Medical School, 2008, vol. 75, no 1, pp. 4–14.

[9] Krabben, T., et al. A feasibility study of the effect of multichannel electrical stimulation and gravity compensation on hand function in stroke patients: A pilot study. In Rehabilitation Robotics (ICORR), 2013 IEEE International Conference on IEEE, 2013. pp. 1–6.

[10] Kim, Y., Chung, Y., Lee, Y.S. The effects of task-oriented electromyography-triggered stimulation on the hand functions of subacute stroke patients. IEEE 19th International Functional Electrical Stimulation Society Annual Conference (IFESS) 2014, IEEE, 2014. pp. 1–4.

[11] Knutson, J.S., et al. Contralaterally controlled functional electrical stimulation for stroke rehabilitation. Annual International Conference of the IEEE Engineering in Medicine and Biology Society. IEEE Engineering in Medicine and Biology Society. Conference. NIH Public Access, 2012. p. 314.

[12] Fischer, H.C., et al. Use of a portable assistive glove to facilitate rehabilitation in stroke survivors with severe hand impairment. IEEE Transactions on Neural Systems and Rehabilitation Engineering, 2016, vol. 24, no. 3, pp. 344–351.

[13] Makowski, N.S., et al. Neuromuscular electrical stimulation to augment reach and hand opening after stroke. Annual International Conference of the IEEE Engineering in Medicine and Biology Society. IEEE, 2011. pp. 3055–3058.

[14] Drop Foot Walking Pattern, Available from: http://www.opnews.com/2014/08/neuro-rehabilitation-how-technological-advancements-are-helping drop-foot-sufferers/7240, [Accessed: 2016-12-23].

[15] Rao, N., et al. The effects of two different ankle-foot orthoses on gait of patients with acute hemiparetic cerebrovascular accident. Rehabilitation Research and Practice, 2014, vol. 2014, p. 301469.

[16] Brunner, R., Meier, G., Ruepp, T. Comparison of a stiff and a spring-type ankle-foot orthosis to improve gait in spastic hemiplegic children. Journal of Pediatric Orthopaedics, 1998, vol. 18, no. 6, p. 719–726.

[17] Custom Orthotic. Ankle foot orthoses (AFO). Available from: http://www.customorthotic.ca/ankle-foot-orthoses. [Accessed: 2016-12-21].

[18] Blaya, J.A., Herr, H. Adaptive control of a variable-impedance ankle-foot orthosis to assist drop-foot gait. IEEE Transactions on Neural Systems and Rehabilitation Engineering, 2004, vol. 12, no. 1, pp. 24–31.

[19] Liberson, W.T. Functional electrotherapy in stimulation of peroneal nerve synchronized with swing phase of gait in hemiparetic patients. Archives of Physical Medicine and Rehabilitation, 1961, vol. 42, pp. 101–105.

[20] Taylor, P., Burridge, J., Dunkerley, A., Wood, D., Norton, J., Singleton, C. Swain, I. Clinical audit of 5 years provision of the odstock dropped foot stimulator. Artificial Organs, 1999, vol. 23, no 5, pp. 440–442.

[21] Odstock System, Available from: http://www.odstockmedical.com, [Accessed: 2016-12-23].

[22] Walkaid System, Available from: http://www.walkaide.com/Pages/default.aspx, [Accessed: 2016-12-23].

[23] MyGait System, Available from: http://www.ottobock.co.uk/neurorehabilitation/product-solutions/mygait/#video,[Accessed: 2016-12-23].

[24] Ness L300 System, Available from: http://www.bioness.com/Products/L300_for_Foot_Drop.php, [Accessed: 2016-12-23].

[25] Foley, N., Peireira, S., Cotoi, A., Serrato, J., Dukelow, S., Sequeira, K., Teasell, R. Chapter 9: Mobility and the lower extremity. Evidence-Based Review of Stroke Rehabilitation. Canadian Partnership for Stroke Recovery. London, Ontario, Canada. pp. 1–175. 2016. (Available Online at www.ebrsr.com, Accessed: 2016-12-23).

[26] Taylor, P., Mann, G., Jolley, C. Swain, I. Economic justification for the odstock dropped foot stimulator (ODFS). The National Clinical FES Centre, Salisbury District Hospital, Salisbury, Wiltshire, SP2 8BJ, UK. IPSO Meeting, 3rd Nov. 2007.

[27] van Swigchem, R., Weerdesteyn, V., van Duijnhoven, H.J., den Boer, J., Beems, T., Geurts, A.C. Near-normal gait pattern with peroneal electrical stimulation as a neuroprosthesis in the chronic phase of stroke: a case report. Archives of Physical Medicine and Rehabilitation, February 2011, vol. 92, no. 2, pp. 320–324.

[28] Stein, R.B., Rolf, R., Everaert, D.G., Bobet, J., Chong, S. Surface electrical stimulation for foot drop: control aspects and walking performance. Journal of Automatic Control, University of Belgrade, 2008, vol. 18, no. 2, pp. 47–52.

[29] Martin, K.D., Polanski, W., Schackert, G., Sobottka. S.B. New therapeutic option for drop foot with the actigait peroneal nerve stimulator—a technical note. World Neurosurgery, December 2015, vol. 84, no. 6, pp. 2037–2042.

[30] STIMuStep System, Available from: http://finetech-medical.co.uk/en-gb/products/stimustep®droppedfootsystem.aspx, [Accessed: 2016-12-23].

[31] Actigait System, Available from: http://www.neurodan.com/products/actigait.aspx, [Accessed: 2016-12-23].

[32] Ottink, A.I., Oostendorp, L.J., Buurke, J.H., Nene, A. V.H., Hermens, J., Ijzerman, M.J. The orthotic effect of functional electrical stimulation on the improvement of walking in stroke patients with a dropped foot: a systematic review. Artif Organs, June 2004, vol. 28, no. 6, pp. 577–86.

[33] Burridge, J., Haugland, M., Larsen, B., Pickering, Niels Svaneborg, R.M., Iversen, H.K., Brøgger Christensen, P., Haase, J., Brennum, J., Sinkjaer, T. Phase II trial to evaluate the ActiGait implanted drop-foot stimulator in established hemiplegia. Journal of Rehabilitation Medicine, 2007, vol. 39, pp. 212–218.

[34] Chang, W.H., Kim, Y-H. Robot-assisted therapy in stroke rehabilitation. Journal of Stroke, 2013, vol. 15, no. 3, p. 174.

[35] Hesse, S. Treadmill training with partial body weight support after stroke: a review. NeuroRehabilitation, 2008, vol. 23, no. 1, pp. 55–65.

[36] Hesse, S., et al. Robot-assisted practice of gait and stair climbing in nonambulatory stroke patients. Journal of Rehabilitation Research and Development, 2012, vol. 49, no. 4, p. 613.

[37] Béjot, Y., Baily, M., Durier, J., Giroud, M. Epidemiology of stroke in Europe and trends for the 21st century. Presse Med. (2016), http://dx.doi.org/10.1016/j.lpm.2016.10.003.

[38] Winstein, C. et al. Guidelines for adult stroke rehabilitation and recovery: a guideline for healthcare professionals from the American heart association/American stroke association. Stroke, 2016, vol. 47, pp. e98–e169.

[39] Van Dokkum, L. Brain computer interfaces for neurorehabilitation – its currents status as a rehabilitation strategy post-stroke. Annals of Physical and Rehabilitation Medicine, 2015, vol. 58, pp. 3–8.

[40] Langhorne, P., Bernhardt, J., Kwakkel, G. Stroke care 2, stroke rehabilitation. Lancet, 2011, vol. 377, pp. 1693–702.

[41] Ward, N.S., Kelly, K., Brander, F. The future of stroke rehabilitation: upper limb recovery. ACNR, 2015, vol. 15, no 4, pp. 6–8.

[42] Perry J. Gait Analysis, Normal and Pathological Function. Slack 1992, Chapter 10–11. SLACK Incorporated, 6900 Grove Road, Thorofare, NJ 08086, USA.

[43] SmartFES System, Available from: http://www.smartfes.cl [Accessed: 2016-12-23].

Chapter 8

Stem Cell Therapy in Pediatric Neurological Disabilities

Alok Sharma, Hemangi Sane,
Nandini Gokulchandran, Prerna Badhe,
Pooja Kulkarni, Suhasini Pai, Ritu Varghese and
Amruta Paranjape

Additional information is available at the end of the chapter

http://dx.doi.org/10.5772/67656

Abstract

Pediatric neurological disorders represent a major part of the disabilities worldwide. In over 10 decades of research to find a cure for these disorders, medical science has not been able to repair the underlying brain injury. This chapter focuses on recent advances in the application of stem cells as a therapeutic tool for some of the common neurodevelopmental disorders (cerebral palsy, autism, intellectual disability and muscular dystrophy). The mechanism of action of stem cells in each disorder has been explained. A review of clinical data has been described giving a clear understanding of current status of stem cell therapy in these disorders. Various factors influencing the outcome of stem cell therapy such as different types of cells, different routes of administration and dosage and frequency of transplantation have also been discussed. Our experience of treating these disorders is exhibited in the form of our published data. Use of novel monitoring tools such as MRI MSK and PET-CT scan brain to track the changes occurring at cellular level after stem cell therapy are described. We also highlight the importance of a multidisciplinary approach of combining rehabilitation with stem cell therapy.

Keywords: stem cell therapy, autism, cerebral palsy, muscular dystrophy, intellectual disability

1. Introduction

Neurodevelopmental disorders (NDD) are characterized by an abnormal development of the brain during the early development phase, leading to a myriad of symptoms and diseases, including delayed milestones and deficits in personal and social functioning [1]. The developmental deficits can vary from specific limitations of adaptive, behavioral and cognitive functioning, motor dysfunction, to global impairments of social skills [2]. Some of the common neurodevelopmental disorders are cerebral palsy (CP), autism spectrum disorders, attention deficit hyperactivity disorder, intellectual disability (ID) or intellectual and developmental disability (IDD), learning disabilities, muscular dystrophies, Down's syndrome, genetic disorders such as fragile-X syndrome, spinal muscular atrophy (SMA) and metabolic disorders.

Pediatric neurological disorders represent a major part of the disabilities worldwide. In over 10 decades of research to find a cure for these disorders, medical science has not been able to repair the underlying brain injury [3]. The causes of NDD can be classified as congenital (present at birth) or acquired (developed after birth). The various etiologies are genetic defects, metabolic disorders, nutritional deficiencies, exposure to toxins, infections, hypoxia/asphyxia, low birth weight, perinatal complications leading to traumatic brain injury or spinal cord injury in children [4]. This may affect language and speech, motor skills, behavior, memory, learning or other neurological functions affecting activities of daily life. While the severity of symptoms often change or evolve as the child's age progresses, these disabilities remain permanent. As these are lifelong disabilities, they pose a substantial economic burden on the society [5]. Hence, finding a treatment for them is the need of the hour. Improvement in the performance of these children would be of great significance to the quality of life of patients and their families.

2. Unmet medical needs

Therapeutic strategies and clinical expectations of patients and medical professionals have not yet been met. Currently, available treatments such as physiotherapy, occupational therapy, behavioral therapy, psychological intervention, speech therapy and pharmacological intervention only focus on alleviating the symptoms of these disabilities and do not address the underlying neuropathophysiology. However, the advent of stem cell therapy has opened new avenues for treatment of pediatric neurological disorders. In recent years, extensive research has been done to explore the potential of stem cells for the treatment of pediatric neurological disabilities. Until now, it was believed that once injured, the cells of the central nervous system cannot regenerate. However, owing to the distinct properties of stem cells to repair and regenerate, they can be considered as a potential therapeutic strategy.

This chapter focuses on recent advances in the application of stem cells as a therapeutic tool for some of the common NDDs (cerebral palsy, autism, intellectual disability and muscular dystrophy). The mechanism of action of stem cells in each disorder has been explained. A review of clinical data has been described giving a clear understanding of current status of stem cell therapy in these disorders. Various factors influencing the outcome of stem cell therapy such as different types of cells, different routes of administration and dosage and frequency of transplantation have also been discussed. Our experience of treating these disorders is exhibited in the form of our published data. Use of novel monitoring tools such as MRI MSK and PET-CT scan brain to track the changes occurring at cellular level after stem cell therapy is described. We also highlight the importance of a multidisciplinary approach of combining rehabilitation with stem cell therapy. Adverse effects of stem cell therapy are also enumerated.

3. What are stem cells?

Stem cells are blank, immature cells which have a capacity to self-renew and differentiate into host-specific multiple lineage cells [6]. Several types of stem cells are being explored for the treatment of neurological disorders such as bone marrow stem cells, embryonic stem cells, olfactory ensheathing cells and umbilical cord blood cells. The main aim of stem cell therapy is replacement of injured/dead neuronal cells and recovery of lost functions [7]. These cells perform repair process directly by regeneration of new cells or indirectly through paracrine activity. The chief underlying mechanisms of stem cells include neuroregeneration, neuroreplacement, neuroprotection, immunomodulation, axon sprouting and neural circuit reconstruction [8] (**Figure 1**).

4. Mechanism of action of stem cells in pediatric neurological disorders

Pediatric neurological disorders are caused due to mechanisms affecting the molecular, cellular and tissue plasticity of the brain and nervous system [9].

Stem cells when transplanted migrate and home towards the injured areas of the brain [10]. This homing property is attributed to the expression of growth factors, chemokine and extracellular matrix receptors on the surface of cells such as stromal cell–derived factor 1 (SDF-1), monocyte chemo attractant protein-3 (MCP-3), stem cell factor (SCF) and/or IL-8. They differentiate into the host tissue cells and replace the injured/dead neuronal tissue [11]. Through paracrine mechanisms they halt further injury and stimulate endogenous cells to carry out the repair and restoration process [12]. Stem cells secrete a vast array of neuroprotective growth factors including brain-derived neurotrophic factor (BDNF), nerve growth factor (NGF), neurotrophin-3 (NT-3), glial cell line–derived neurotrophic factor (GDNF) and insulin-like

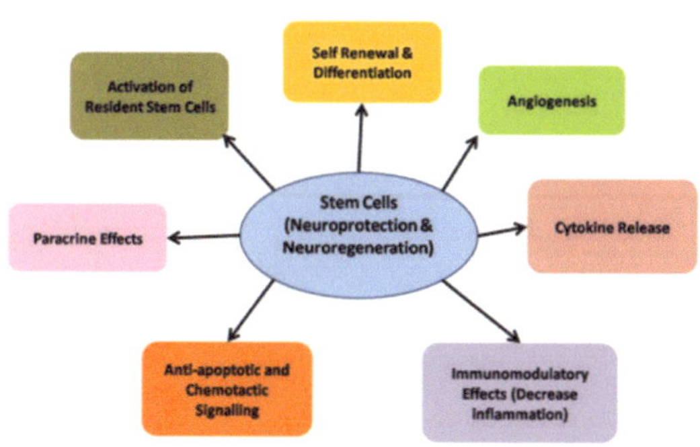

Figure 1. Mechanism of action of stem cells in pediatric neurological disorders.

growth factor type 1. These growth factors activate a number of signaling pathways and help in enhancing differentiation, survival of neurons and maintaining neuronal functions [13]. They also produce vascular endothelial growth factor (VEGF), hepatocyte growth factor (HGF) and basic fibroblast growth factor (FGF-2) which improve perfusion and enhance angiogenesis [14]. Anti-inflammatory paracrine factors such as Interleukin 10 (IL 10) and Transforming growth factor (TGF)-β help in immunomodulation [15].

5. Clinical application of stem cell therapy in NDD

In this section, we discuss the literature review of various stem cell therapy studies in each disorder followed by our experience.

We published a study of 71 children diagnosed with different incurable neurological disorders. Autologous bone marrow–derived mononuclear cells were transplanted intrathecally and intramuscularly. Improvements were noted in muscle power, functional independent measure (FIM) and Brooke and Vignos scale. Imaging and electrophysiological investigations also showed improvement. Overall 97% muscular dystrophy cases showed subjective, functional and investigational improvement. Eighty-five percent of cases of cerebral palsy cases showed improvements. Eighty-eight percent of cases of other incurable neurological disorders such as autism, Retts syndrome and giant axonal neuropathy also showed improvement. No major adverse events were noted.

6. Stem cell therapy in cerebral palsy

In cerebral palsy, white matter injury also known as periventricular leukomalacia (PVL) is one of the major pathologies observed [16]. Stem cells differentiate into neurons, oligodendrocytes and astrocytes which replace and repair the white matter injury in CP [17] (**Figure 2**). The growth factors secreted by these cells also help in remyelination, synaptogenesis, cytoprotection and angiogenesis which reverse the cellular injury in CP [18, 19]. Numerous preclinical studies have

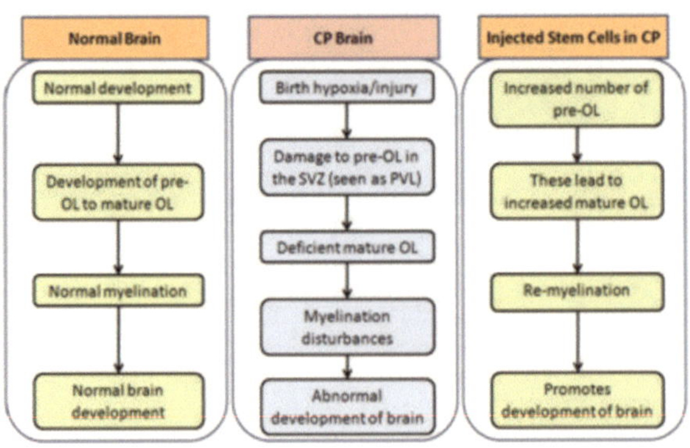

Figure 2. Stem cell therapy in cerebral palsy.

http://dx.doi.org/10.5772/67656

demonstrated the potential of stem cell transplantation in cerebral palsy. The homing property of these cells was confirmed by Chen et al., who transplanted magnetically labeled mesenchymal stem cells in a model of perinatal brain injury and found that these cells migrate to lesion sites and proliferate [20]. Studies have demonstrated the differentiation of bone marrow, umbilical cord blood, neural and other progenitor stem cells into neurons and oligodendrocytes in experimental animal models [21–25]. Transplantation of stem cells in rat models have resulted in improved cognition and sensorimotor deficits along with functional recovery [26].

6.1. Clinical evidence

In cerebral palsy, around 26 studies have been published explaining the effect of stem cell therapy. Overall, 579 (90%) out of 646 patients have shown improvements (**Table 1**) [20, 27–51].

Sr. no	Citations	Cells used	Route of administration	Sample size	Patient improved	Adverse events
1.	Sharma et al. [27]	Autologous bone marrow mononuclear cells (BMMNCs)	Intrathecal	40	None	
2.	Min et al. [28]	Allogenic umbilical cord blood	Intravenous	96	86	Pneumonia and irritability
3.	Lee et al. [29]	Autologous umbilical cord blood	Intravenous	20	5	Nausea, hemoglobinuria or urticaria
4.	Purandare et al. [30]	Autologous BMMNCs	Intrathecal	1	1	None
5.	Chen et al. [20]	Autologous bone marrow mesenchymal cells	Subarachnoid	60	60	Increased frequency of crying
6.	Li et al. [31]	Autologous bone marrow mesenchymal cells	Subarachnoid	1	1	None
7.	Luan et al. [32]	Neural progenitor cells	Intracranial	45	45	None
8.	Chen et al. [33]	Olfactory ensheathing cells	Intracranial	33	33	None
9.	Ramirez et al. [34]	Umbilical cord blood cells	Intramuscular injection	8	8	Localized mild pain at the site of injection.
10.	Payne [35]	Umbilical cord blood cells	Subcutaneous	16	16	None
11.	Sharma et al. [36]	Autologous BMMNCs	Intrathecal	1	1	None
12.	Sharma et al. [37]	Autologous BMMNCs	Intrathecal	1	1	None
13.	Papadopoulos et al. [38]	Autologous BMMNCs	Intrathecal	2	2	None
14.	Sharma et al. [39]	Autologous BMMNCs	Intrathecal	1	1	None
15.	Jensen and Hamelmann [40]	Autologous umbilical cord blood cells	Intravenous	1	1	None
16.	Wang et al. [41]	Umbilical cord mesenchymal stem cells	Intravenous and intrathecal administration	1	1	Temporary low-grade fever
17.	Luan et al. [42]	Human neural stem cells	Intracerebral	7	4	None

Sr. no	Citations	Cells used	Route of administration	Sample size	Patient improved	Adverse events
18.	Wang et al. [43]	Bone marrow mesenchymal stromal cells	–	52	52	none
19.	Yang et al. [44]	Umbilical cord mesenchymal stem cell	Intravenous and intrathecal	25	22	none
20.	Zali et al. [45]	CD133-positive enriched bone marrow progenitor cells	Intrathecal	12	12	seizure
21.	Mancías-Guerra et al. [46]	Autologous bone marrow–derived total nucleated cell (TNC)	Intrathecal and intravenous injection	18	18	Headache, vomiting, fever and stiff neck
22.	Romanov et al. [47]	Allogenic umbilical cord blood cells	Intravenous	80	80	None
23.	Zang et al. [48]	Umbilical cord blood mesenchymal stem cells	Intravenous	1	1	None
24.	Wang et al. [49]	Umbilical cord–derived mesenchymal stromal cell	Subarachnoid	16 (8 pair of twins)	16 (8 pair of twins)	None
25.	Shroff et al. [50]	Human embryonic stem cells	Intravenous	91	63	Seizures
26.	Abi Chahine et al. [51]	Bone marrow mononuclear cells	Intrathecal	17	11	Headaches, transient fever and vomiting

Table 1. Clinical evidence demonstrating the use of stem cells in cerebral palsy.

In 2015, we published a nonrandomized study demonstrating the benefits of autologous bone marrow mononuclear cells (BMMNCs) in cerebral palsy [27]. These patients were followed up at 3 and 6 months. Six months after intervention, 38 out of 40 (95%) patients showed improvements and 2 did not show any improvement but remained stable without any deterioration (**Figure 3**). No major adverse events were noted except for seizures in two patients which were controlled by medications.

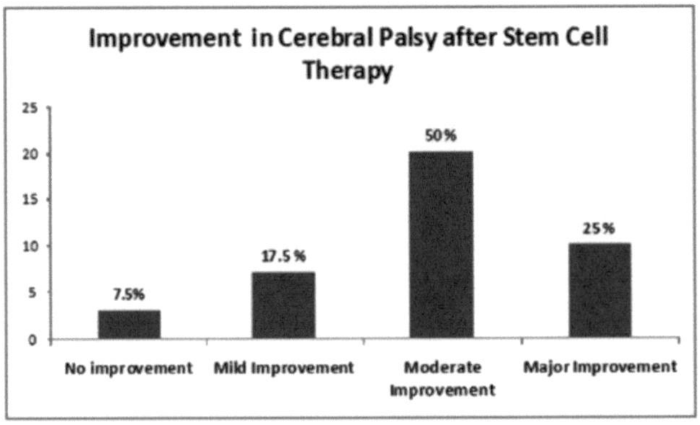

Figure 3. Graph showing improvement in children with cerebral palsy after stem cell therapy.

We have also published three case reports demonstrating the safety and efficacy of BMMNC transplantation in cerebral palsy [36, 37, 39]. In these case reports, the functional improvements are supported by improved brain metabolism recorded in comparative PET-CT scans performed before and after the intervention.

7. Stem cell therapy in autism

In autism, immune dysfunction, hypoperfusion, oxidative stress, decreased number of Purkinje cells (PCs), cerebellum alterations, defective cortical organization and altered plasticity of dendritic spine morphology are the underlying neuropathologies (**Figure 4**) [52, 53]. Stem cells modulate the immune dysfunction by releasing anti-inflammatory molecules and inhibiting pro-inflammatory molecules, which further reduces neural injury [54]. They also facilitate angiogenesis which increases blood and oxygen supply to the brain thus reversing the hypoperfusion [55]. Stem cells may also reinforce cortical plasticity, promote synaptic plasticity and restore cerebellar PCs [56]. These mechanisms collectively may improve the lost neural connectivity and restore lost functions in autism. In an experimental model of mice, H Segal Gavish et al. transplanted mesenchymal stem cells, which resulted in reduction of stereotypical behaviors, decrease in cognitive rigidity and improvement in social behavior. Tissue analysis revealed elevated BDNF protein levels in the hippocampus accompanied by increased hippocampal neurogenesis in the MSC-transplanted mice compared with sham treated mice [57].

7.1. Clinical evidence

A total of 11 studies (3 case series and 8 case reports) have been published all over the world demonstrating the benefits of stem cell therapy in autism. Overall, 122 patients were administered with cellular therapy and 90 showed improvements (**Table 2**) [58–68].

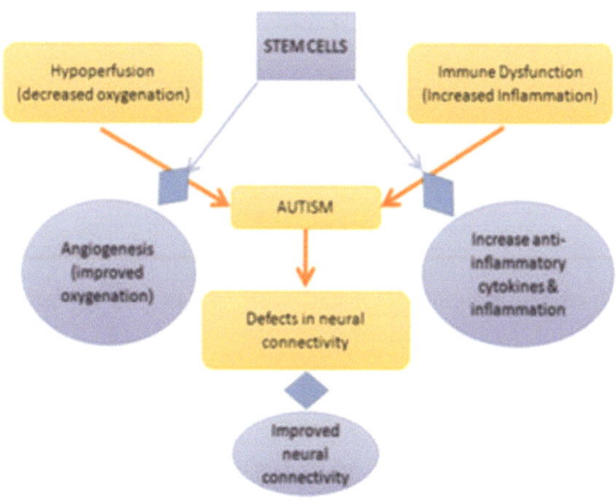

Figure 4. Stem cell therapy in autism.

Author	Type of cells used	Route of administration	Sample size	How many patients improved	Demonstrated safety
Sharma et al. [58]	Autologous bone marrow mononuclear cells (BMMNCs)	Intrathecal	32	29	Yes
Lv et al. [59]	Human cord blood mononuclear cells (CBMNCs) and umbilical cord–derived mesenchymal stem cells (UCMSCs)	Intravenous and intrathecal	37	18	Yes
Bradstreet et al. [60]	Fetal stem cells	Subcutaneous	45	35	Yes
Sharma et al. [61]	Autologous BMMNCs	Intrathecal	1	1	Yes
Sharma et al. [62]	Autologous BMMNCs	Intrathecal	1	1	Yes
Sharma et al. [63]	Autologous BMMNCs	Intrathecal	1	1	Yes
Sharma et al. [64]	Autologous BMMNCs	Intrathecal	1	1	Yes
Sharma et al. [65]	Autologous BMMNCs	Intrathecal	1	1	Yes
Sharma et al. [66]	Autologous BMMNCs	Intrathecal	1	1	Yes
Sharma et al. [67]	Autologous BMMNCs	Intrathecal	1	1	Yes
Sharma et al. [68]	Autologous BMMNCs	Intrathecal	1	1	Yes

Table 2. Clinical evidence demonstrating the use of stem cells in autism.

In 2013, we published an open label proof of concept study which included 32 patients of autism [58] (**Figure 5**). These patients were followed up for 26 months (mean 12.7). The outcome measures used were Childhood Autism Rating Scale (CARS), Indian Scale for Autism Assessment (ISAA), Clinical Global Impression (CGI) and Functional Independence Measure (FIM/Wee-FIM) scales. It was found that out of 32 patients, a total of 29 (91%) patients improved on total ISAA

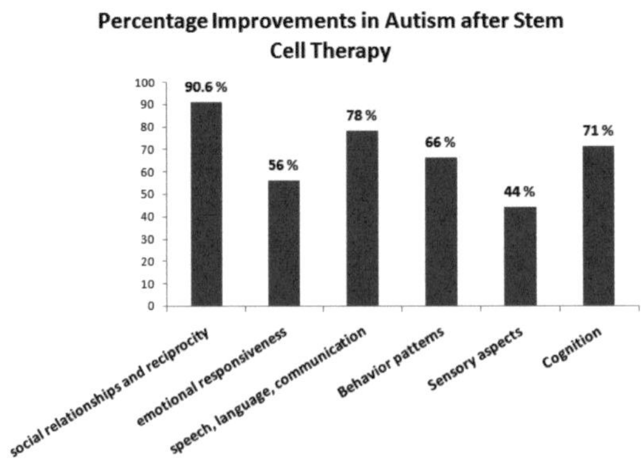

Figure 5. Graph showing percentage improvement in various symptoms of autism post stem cell therapy.

scores and 20 patients (62%) showed decreased severity on CGI-I. On CGI-II 96% of patients showed global improvement. Improvements in brain metabolism were also observed on positron emission tomography-computed tomography (PET-CT) scan brain. All 32 patients were monitored through the duration of follow-up for any major adverse events. Incidence of seizures was recorded in three patients, which were reversible and easily controlled with medications.

In addition to the above study, we have also published eight case reports demonstrating the safety, efficacy and objective improvements on PET-CT scan brain in patients with autism following stem cell therapy [61–68].

8. Stem cell therapy in intellectual disability

In intellectual disability (ID), the neuronal connectivity in the brain is impaired along with disrupted cell migration, cell multiplication, axon growth, brain plasticity and synaptogenesis (**Figure 6**) [9]. Studies have recorded defects in hippocampus and cerebral cortex areas of the brain leading to faulty information processing, consecutively affecting cognition and adaptive behavior in ID. Stem cells restore the synaptic transmitters released and provide local reinnervations to the area affected. It also integrates existing neural and synaptic network and re-establishes connections of functional afferent and efferent cells which may have contributed in restoring the cognitive and functional deficit in IDs [69].

8.1. Clinical evidence

We are currently under process of analyzing the data of a prospective study conducted to demonstrate the effect of autologous bone marrow mononuclear cells in intellectual disability.

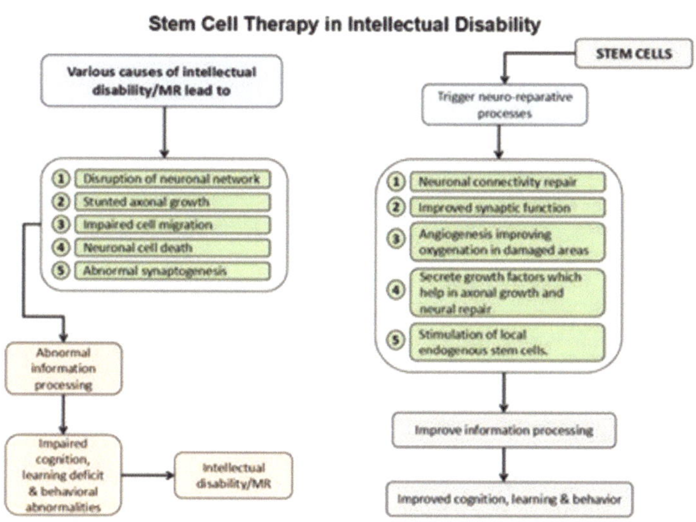

Figure 6. Stem cell therapy in intellectual disabilities.

However, in 2015, we published a report of a 13-year-old boy with intellectual disability who exhibited improvements after stem cell therapy [70]. He was followed up after 3 and 6 months of intervention. No major adverse events were recorded post intervention. Over a period of 6 months, he showed improved eye contact, cognition, learning ability, behavior and ability to perform activities of daily living. His score on Functional Independence Measure (FIM) increased from 67 to 76. On comparing the pre and post PET-CT scan, improvement in metabolic activity of hippocampus, left amygdala and cerebellum was recorded. These changes correlated to the functional outcome.

9. Stem cell therapy in Duchenne muscular dystrophy

The underlying pathogenic mechanism of muscular dystrophy is an imbalance between muscle degeneration and resident satellite cell–mediated regeneration [71]. Satellite cells, the adult skeletal muscle progenitor cells, are considered to be the main cell type involved in skeletal muscle regeneration. Continuous cycles of degeneration and regeneration of muscle fibers exhausts the muscle stem cell pool, leading to muscle being replaced by adipose and fibrotic tissue. Stem cell therapy holds great promise as a treatment for Duchenne muscular dystrophy by providing cells that can both deliver functional muscle proteins and replenish the stem cell pool [72].

Stem cells are known to enhance angiogenesis, contribute to neovascularization, promote tissue remodeling, prevent apoptosis, decrease inflammation, release growth factors and activate the satellite cells [73] (**Figure 7**). In animal models, these cells have shown to produce the deficient proteins and make new muscle cells which fuse with the host fibers. Further, stem cell–derived exosomes which are small membrane vesicles and are responsible for inter-cellular communication, promote muscle regeneration by enhancing myogenesis and angiogenesis [74].

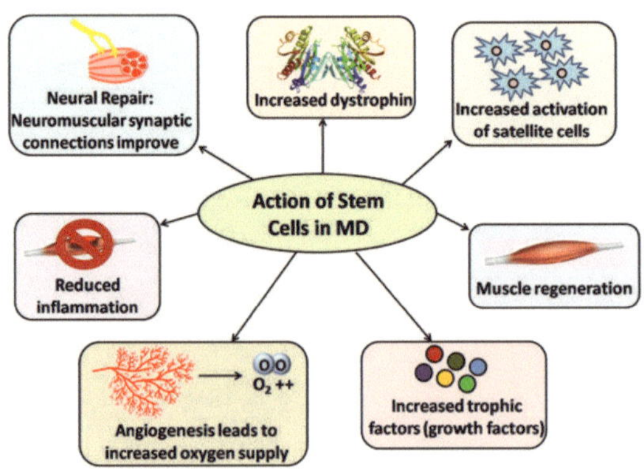

Figure 7. Role of stem cells in muscular dystrophy.

9.1. Clinical evidence

A total of 14 studies have been conducted demonstrating the efficacy of stem cells in muscular dystrophy. Various types of stem cells such as bone marrow–derived cells, umbilical cord stem cells and muscle-derived cells were used. Out of a total of 346 patients who underwent stem cell therapy, 296 showed a positive outcome (**Table 3**) [75–90].

Author	Sample size	Type of cells used	Route of administration	Number of patients improved	Level of evidence
Torrente et al. [75]	8	Muscle-derived CD133+ cell	Intramuscular	8	4
Yang et al. [76]	82	Autologous bone marrow mesenchymal stem cells (BMSC) and umbilical cord mesenchymal stem cells (UMSC)	Intravenous and intramuscular	Effective in 68 [82.9%] cases.	4
Mendell et al. [77]	12	Muscle precursor cells	Intramuscular	In one patient, 10.3% of muscle fibers expressed donor-derived dystrophin after myoblast transfer. Three other patients also had a low level of donor dystrophin; eight had none.	4
Sharma et al [78]	150	BMMNCs	Intrathecal, Intramuscular	130 [86.67%] cases showed symptomatic and functional improvements	4
Rajput et al. [79]	16	Human umbilical cord mesenchymal stem cells	IV and IM injection	9 out of 11 patients were stable no deterioration.	
Sharma et al. [80]	65	BMMNCs	Intrathecal, Intramuscular	65 (plateau phase, no further progression)	4
Skuk et al. [81]	1	Muscle-precursor cells	Intramuscular	27.5% of the myofiber profiles expressed donor-derived dystrophin, 1 month post-transplantation and 34.5%, 18 months post-transplantation	5
Sharma et al. [82–87]	6 case reports	BMMNCs	Intrathecal, Intramuscular	6	5
Kang et al. [88]	1	Umbilical cord–derived hematopoietic stem cell	Intrathecal	not effective	5
Skuk et al. [89]	3	Myogenic cells	Intramuscular	dystrophin-positive myofibers in the cell-grafted sites amounting to 9 (patient 1), 6.8 (patient 2) and 11% (patient 3).	
Zhang et al. [90]	1	Allogeneic cord blood stem cells	Intravenous	1	5

Table 3. Clinical evidence demonstrating the use of stem cells in muscular dystrophy.

We conducted a study on 150 patients diagnosed with muscular dystrophy. On a mean follow up period of 12 months ± 1 month, 86.67% cases showed symptomatic and functional improvements, with six patients showing muscle regeneration and decrease in fatty infiltration on musculoskeletal magnetic resonance imaging (MRI MSK) and nine showing improved muscle electrical activity on electromyography (EMG). Fifty-three percent cases showed increase in trunk muscle strength, 48% an increase in upper limb (UL) strength, 59% an increase in lower limb (LL) strength and about 10% showed an improved gait pattern (**Figures 8** and **9**).

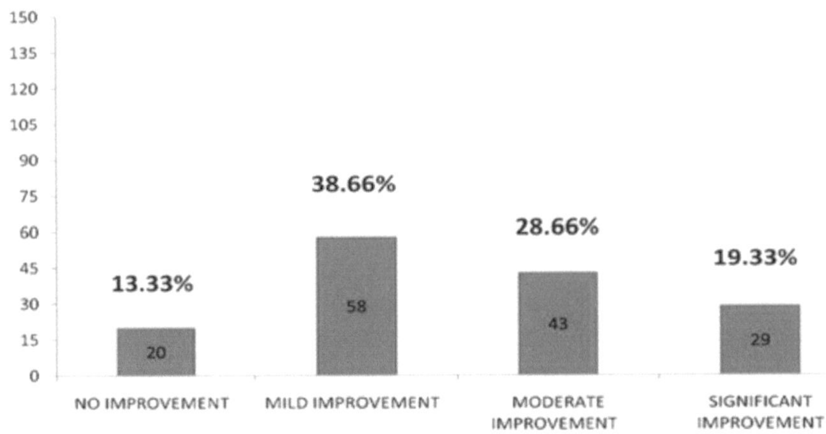

Figure 8. Graph showing improvements in muscular dystrophy patients after stem cell therapy. *y*-axis = number of patients ($n = 150$).

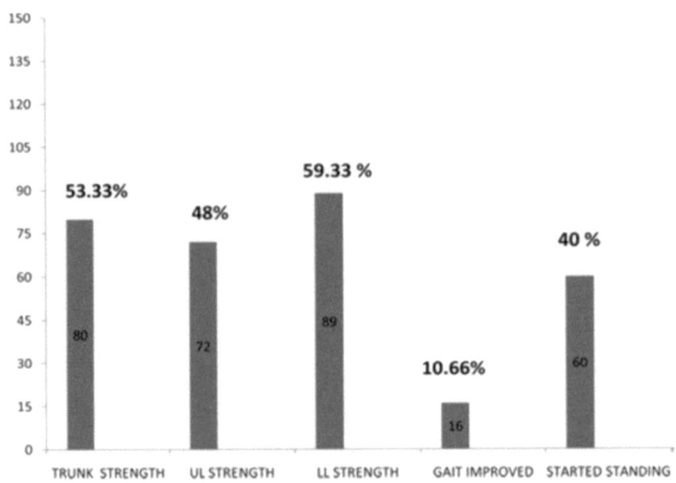

Figure 9. Graph showing symptomatic improvements in muscular dystrophy patients after stem cell therapy. Number of patients showing improvements in trunk strength, upper limb (UL) strength, lower limb (LL) strength, gait pattern, and standing function are shown. *y*-axis = number of patients ($n = 150$).

10. Adverse events of stem cell therapy

The adverse events following stem cell transplantation mainly depends on the type of stem cell and the route of administration. Other factors like dosage of cells, frequency of transplantation and age of the patient may also contribute. Fetal stem cells are known to be potentially tumorigenic [91]. Use of umbilical cord stem cells is limited due to slow or incomplete immune reconstitution, resulting in a high transplantation-related mortality (TRM) due to infections. Most studies have demonstrated a predominance of Gram-positive bacteria (GPB) bloodstream infections [92]. On the contrary, adult stem cell has not shown any serious adverse events. Autologous cell transplantation is safer than allogenic.

Adverse events of stem cell therapy can be categorized into minor and major adverse events. Minor adverse events include procedure related events such as spinal headache, nausea, diarrhea, vomiting, pain or bleeding at the site of aspiration/injection and fever amongst others. These are treated using medications. Anesthetic complications and allergic reactions may also occur depending on the procedure. Major adverse events include episodes of seizures occurring after intervention. These can be managed prophylactically. Pre-existing epileptogenic focus in Electroencephalogram (EEG) also predicts the occurrence of seizures. Evidence suggests that antiepileptic prophylactic regimen decreases the incidence of seizures as an adverse event after stem cell therapy [93].

11. Factors influencing the outcome of stem cell therapy for NDDs

11.1. Routes of administration

The route of delivery of cells plays an important role in maximizing the clinical output of cellular therapy. Intrathecal route of administration is a relatively minimally invasive and targeted route of administration of cells. It is devoid of any major side effects [94]. In neurological disorders, intrathecal transplantation enhances the accessibility of the injected cells into the CNS [95]. Intramuscular injections are administered at the motor points plotted on the affected muscles. Motor points are the points where the innervating nerve enters the muscle. Thus, implantation of cells in the muscles enhances the effect of stem cells on the degenerating muscles [96]. Intravenous administration is the least invasive route. However, evidence suggests that majority of cells get trapped in the pulmonary passage and only few cells reach the injured site [97]. An alternate route of administration is via intra-cerebral route. But, it is an invasive technique and might result in secondary complications such as bleeding and neural tissue injury [98]. Hence, as compared to all the delivery routes, intrathecal administration is most efficacious.

11.2. Types of cells

Cells used from allogenic sources have an inherent risk of immunogenicity and may potentially cause immune rejection of graft versus host disease. Autologous cells have the least possibility of immune reaction and so far clinical studies with autologous minimally manipulated cells have shown no immunogenic reactions in the host post transplantation. Autologous cells may therefore be a safer option in children with NDD.

11.3. Etiology

Genetic factors play a major part in the pathology of neurological disorders and gene therapy has provided novel insights in treating the underlying genetic aberrations. But gene therapy cannot replace the lost neurons and practical difficulties have prevented it from being a clinically feasible and viable option at present. The sporadic nature of the disease is also an important factor influencing the outcome, where the etiology of the disease is unknown. Stem cell therapy addresses the core injury occurring in the brain. The multiple mechanism of action of the stem cells addresses the multifactorial pathology of the NDD.

11.4. Severity

It has been observed that the mild cases of neurodevelopment disorders have a better recovery curve than the chronic cases. In mild cases, axonal function remains intact and recovery can be rapid if remyelination occurs. In severe cases, axonal degeneration occurs and recovery depends on axonal regeneration. Recovery becomes much slower, and there is a greater degree of residual injury. Mild cases require lesser dosage of cells and the frequency of doses required is less to attain potential recovery than the severe cases.

11.5. Age of the patient

One of postulated hypothesis is that the neural circuits, that form the basis for learning, behavior and health, are more plastic during the initial years of life. They become increasingly difficult to alter over time. Age-related decline in the potency of the stem cells is observed which might also affect the remodeling of CNS by these cells. Early intervention is advised for better outcome of stem cell therapy.

12. Importance of neurorehabilitation

Neurorehabilitation aims at restoration and maximization of functions that have been lost due to impairments caused by injury or disease of nervous system making the patient functionally independent. The rehabilitation regime promotes and facilitates neural plasticity [99]. Studies have shown that exercise enhances the effect of injected stem cells by inducing mobility of the cells, activating and proliferating the local stem cells, promoting muscle angiogenesis and release of cytokines and nerve growth factors. Hence, neurorehabilitation compliments with the stem cell therapy [100].

http://dx.doi.org/10.5772/67656

13. Objective evidence: neuroimaging techniques to monitor the outcome of stem cell therapy

Neuroimaging techniques enable the quantitative measurement of various biological markers which may serve as a powerful tool for optimizing the use of stem cells for clinical applications.

13.1. Positron emission tomography (PET-CT) scan brain

PET-CT scan brain can be used efficiently as a monitoring tool to study the outcome of stem cell therapy. One of the advantages of using PET-CT is its extreme sensitivity enabling it to detect molecules at the nanomolar level [101]. Brain 18F-FDG PET allows studying the cerebral glucose metabolism, indicating the neuronal and synaptic activity. It dynamically measures the energy metabolism along with blood oxygenation and blood flow [102]. The alteration in neuronal activity caused by disease is reflected in change of glucose metabolism and can be revealed in the PET-CT scan brain. As mentioned previously in the clinical results, there were improvements recorded in the brain metabolism of patients included in the clinical studies. The changes seen on PET-CT scan brain correlated with the clinical improvement indicating that it can identify alteration occurring at the tissue levels (**Figures 10–12**).

13.2. Magnetic resonance imaging (MRI)

Magnetic resonance imaging (MRI) is gaining popularity because of its capacity to reveal characteristic findings that address the diagnosis and support therapeutic interventions.

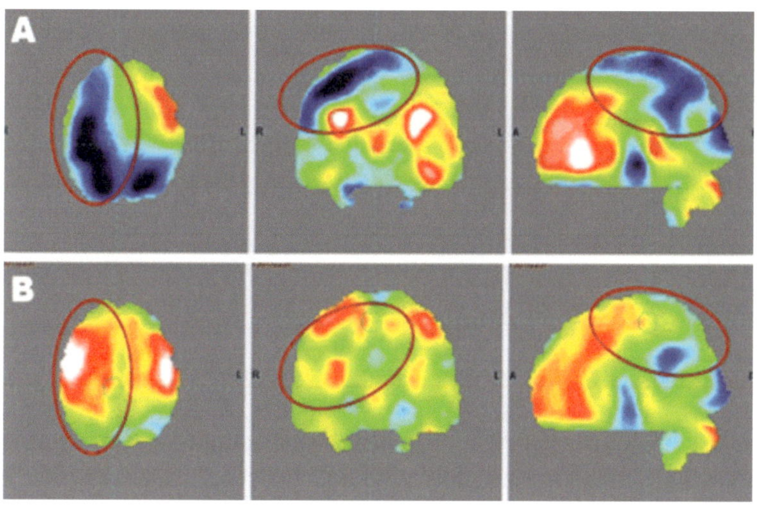

Figure 10. (A) Pre SCT PET-CT scan images with blue areas indicating hypometabolism. (B) These areas have almost disappeared after SCT as seen in the post PET-CT scan image. This shows improvement in the metabolism/functioning in the affected areas of the brain after SCT.

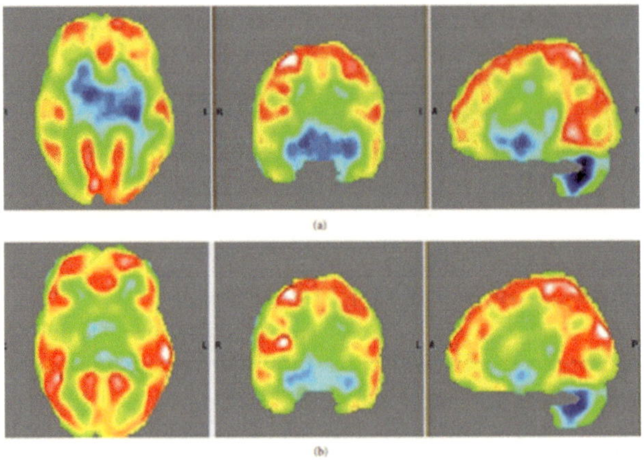

Figure 11. Findings in PET-CT scan before and after cellular therapy. (a) PET-CT scan before intervention showing reduced FDG uptake in the areas of frontal lobe, cerebellum, amygdala, hippocampus, parahippocampus, and mesial temporal lobe. (b) PET-CT scan six months after intervention comparison shows increased FDG uptake in the areas of frontal lobe, cerebellum, amygdala, hippocampus, parahippocampus, and mesial temporal lobe.

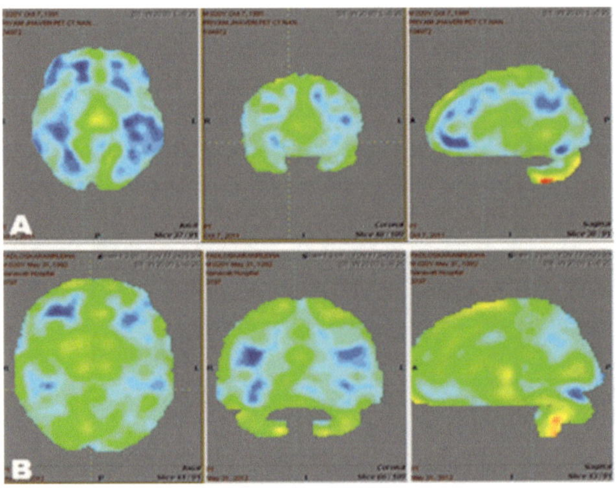

Figure 12. (A) Pre stem cell therapy PET-CT scan showing blue areas with hypometabolism. (B) Post stem cell therapy PET-CT scan showing decrease in blue areas which is replaced by green areas indicating improved functioning of the brain.

Since MRI is devoid of ionizing radiations, it has turned out to be a valuable imaging method in children, although sometimes sedation might be necessary. In the past few years, studies have reported on the detection of muscle involvement pattern in various muscular dystrophies through MRI musculoskeletal imaging (MRI/MSK). The images provide a high soft tissue contrast allowing assessment of affected striated muscles in terms of shape, volume (hypotrophy and hypertrophy) and architecture [103, 104]. MRI MSK was used as a tool to assess the therapeutic efficacy of stem cell therapy in muscular dystrophy. In our published data, images of MRI MSK performed after intervention has revealed stabilization of disease progression in muscular dystrophy.

14. Conclusion

In children, the brain is still at a developing stage and not fully matured resulting in maximal neural plasticity during childhood. Hence, likelihood of improvement in affected areas of the brain increases manifold with early intervention. Stem cell therapy has recently gained lot of importance as a therapeutic strategy for various disorders including NDDs. In this review, we have demonstrated the outcome of stem cell therapy in NDDs mainly cerebral palsy, autism, intellectual disability and muscular dystrophy supported by our published data. Through its neurorestorative and neuroregenerative property, stem cells have the capacity of repairing the underlying neural and muscular dysfunction. This property can augment neurodevelopment, facilitating achievement of milestones earlier as compared to the current conventional treatment modalities. In progressive developmental disorders like muscular dystrophies, stem cell therapy has shown to slow down the disease progression. The data also establishes the fact that autologous stem cell therapy is a safe and efficacious treatment which helps in recovery of lost functions and neural plasticity.

Though stem cell therapy is not a cure, the gap between normalcy and disability can be minimized. Stem cells in combination with the multidisciplinary medical and rehabilitative modalities can enhance and hasten the recovery from NDD which will help the patient to lead a productive and respectable life in the society.

15. Future directions

Stem cell therapy is still in its developing stage. There are still numerous uncertainties prevailing with respect to optimum volume of cells to be injected, number of doses, route of administration, types of cells amongst others. The advent of induced pluripotent stem cells (iPSCs) has provided opportunities for the study of human neurodevelopmental diseases in a controlled environment. Reprogramming cells from patients with neurological diseases will allow the study of disease-specific cellular and molecular pathways causing these diseases. Also, the establishment of neural stem cells (NSCs), a life-long source of neurons and glia, has contradicted the dogma that the nervous system lacked regenerative power. Future studies need to focus on the precautionary pre-intervention assessments to identify patients with high risk for seizures and related adverse events after stem cell therapy. A better knowledge of all these factors will improve the therapeutic effectiveness of stem cell therapy. Future studies should consider the use of modern radiological tools as monitoring tool and substantiate the effects of cellular therapy in NDDs. Large scale, multicentre and randomized controlled trials are recommended to further establish the safety and efficacy of cellular therapy.

Author details

Alok Sharma[1], Hemangi Sane[2], Nandini Gokulchandran[1], Prerna Badhe[3], Pooja Kulkarni[2,3], Suhasini Pai[2], Ritu Varghese[2] and Amruta Paranjape[2,4]

*Address all correspondence to: poojakul28@gmail.com

1 Department of Medical Services and Clinical research, NeuroGen Brain & Spine Institute, Navi Mumbai, India

2 Department of Research & Development, NeuroGen Brain & Spine Institute, Navi Mumbai, India

3 Department of Regenerative Laboratory Services, NeuroGen Brain & Spine Institute, Navi Mumbai, India

4 Department of Neurorehabilitation, NeuroGen Brain & Spine Institute, Navi Mumbai, India

References

[1] Telias M, Ben-Yosef D. Modeling neurodevelopmental disorders using human pluripotent stem cells. Stem Cell Reviews and Reports. 2014;**10**(4):494–511

[2] American Psychiatric Association. Neurodevelopmental Disorders: DSM-5® Selections. Arlington, VA: American Psychiatric Pub; 2015

[3] Mullin AP, Gokhale A, Moreno-De-Luca A, Sanyal S, Waddington JL, Faundez, V. Neurodevelopmental disorders: Mechanisms and boundary definitions from genomes, interactomes and proteomes. Translational Psychiatry. 2013;**3**(12):e329

[4] Kelleher E, Corvin A. Overlapping etiology of neurodevelopmental disorders. In: Mitchell KJ, editor. The Genetics of Neurodevelopmental Disorders. New Jersey: John Wiley & Sons, Inc. 2015.

[5] World Health Organization. Neurological Disorders: Public Health Challenges. Geneva: World Health Organization; 2006.

[6] Ul Hassan A, Hassan G, Rasool Z. Role of stem cells in treatment of neurological disorder. International Journal of Health Sciences. 2009;**3**(2):227–233

[7] Lindvall O, Björklund A. Cell replacement therapy: Helping the brain to repair itself. NeuroRx. 2004;**1**(4):379–381

[8] Karussis D, Kassis I, Kurkalli BG, Slavin S. Immunomodulation and neuroprotection with mesenchymal bone marrow stem cells (MSCs): A proposed treatment for multiple sclerosis and other neuroimmunological/neurodegenerative diseases. Journal of Neurological Science. 2008;**265**(1-2):131–135

[9] Kroon T, Sierksma MC, Meredith RM. Investigating mechanisms underlying neurodevelopmental phenotypes of autistic and intellectual disability disorders: A perspective. Frontiers in Systems Neuroscience. 2013;**7**:75

[10] Smart N, Riley PR. The stem cell movement. Circulation Research. 2008;**102**(10):1155–1168

[11] Verma RS. Breaking dogma for future therapy using stem cell – Where we have reached? The Indian Journal of Medical Research. 2016;**143**(2):129–131

[12] Baraniak PR, McDevitt TC. Stem cell paracrine actions and tissue regeneration. Regenerative Medicine. 2010;**5**(1):121–143

[13] Hung C-W, Liou Y-J, Lu S-W, et al. Stem cell-based neuroprotective and neurorestorative strategies. International Journal of Molecular Sciences. 2010;**11**(5):2039–2055

[14] Crisostomo PR, Wang M, Herring CM, Markel TA, Meldrum KK, Lillemoe KD, et al. Gender differences in injury induced mesenchymal stem cell apoptosis and VEGF, TNF, IL-6 expression: role of the 55 kDa TNF receptor (TNFR1). Journal of Molecular and Cellular Cardioliology. 2007;**42**(1):142–149

[15] Markel TA, Crisostomo PR, Wang M, Herring CM, Meldrum DR. Activation of individual tumor necrosis factor receptors differentially affects stem cell growth factor and cytokine production. American Journal Physiology Gastrointestinal Liver Physiology. 2007;**293**(4):G657–G662

[16] Khwaja O, Volpe JJ. Pathogenesis of cerebral white matter injury of prematurity. Archives of Disease in Childhood Fetal and Neonatal Edition. 2008;**93**(2):F153-F161

[17] Castillo-Melendez M, Yawno T, Jenkin G, Miller SL. Stem cell therapy to protect and repair the developing brain: A review of mechanisms of action of cord blood and amnion epithelial derived cells. Frontiers in Neuroscience. 2013;**7**:194.

[18] Giulian D, Johnson B, Krebs JF, Tapscott MJ, Honda S. A growth factor from neuronal cell lines stimulates myelin protein synthesis in mammalian brain. Journal of Neuroscience. 1991;**11**(2):327–336

[19] Garofalo L, Ribeiro-da-Silva A, Cuello AC. Nerve growth factor-induced synaptogenesis and hypertrophy of cortical cholinergic terminals. Proceeding of National Academy of Sciences of the United States of America. 1992;**89**:2639–2643

[20] Chen G, Wang Y, Xu Z, Fang F, Xu R, Wang Y, Hu X, Fan L, Liu H. Neural stem cell-like cells derived from autologous bone mesenchymal stem cells for the treatment of patients with cerebral palsy. Journal of Translational Medicine. 2013;**11**:21

[21] Woodbury D, Schwarz EJ, Prockop DJ, Black IB. Adult rat and human bone marrow stromal cells differentiate into neurons.Journal of Neuroscience Research. 2000;**61**:364–370

[22] Wang XL, Zhao YS, Hu MY, Sun YQ, Chen YX, Bi XH. Umbilical cord blood cells regulate endogenous neural stem cell proliferation via hedgehog signaling in hypoxic ischemic neonatal rats. Brain Research. 2013;**1518**:26–35

[23] Park KI, Himes BT, Stieg PE, Tessler A, Fischer I, Snyder EY. Neural stem cells may be uniquely suited for combined gene therapy and cell replacement: evidence from engraftment of Neurotrophin-3-expressing stem cells in hypoxic-ischemic brain injury. Experimental Neurology. 2006;**199**(1):179–190.

[24] Yasuhara T, Matsukawa N, Yu G, Xu L, Mays RW, Kovach J, Deans RJ, Hess DC, Carroll JE, Borlongan CV. Behavioral and histological characterization of intrahippocampal grafts of human bone marrow-derived multipotent progenitor cells in neonatal rats with hypoxic-ischemic injury. Cell Transplant. 2006;**15**(3):231–238.

[25] Webber DJ, van Blitterswijk M, Chandran S. Neuroprotective effect of oligodendrocyte precursor cell transplantation in a long-term model of periventricular leukomalacia. American Journal of Pathology. 2009;**175**(6):2332–2342

[26] Pimentel-Coelho PM, Magalhães ES, Lopes LM, deAzevedo LC, Santiago MF, Mendez-Otero R. Human cord blood transplantation in a neonatal rat model of hypoxic-ischemic brain damage: Functional outcome related to neuroprotection in the striatum. Stem Cells and Development. 2010;**19**(3):351–358

[27] Sharma A, Sane H, Gokulchandran N, Kulkarni P, Gandhi S, et al. A clinical study of autologous bone marrow mononuclear cells for cerebral palsy patients: a new frontier. Stem Cells International. 2015;**2015**:11 pages. Article ID 905874

[28] Min K, Song J, Kang JY, Ko J, Ryu JS, Kang MS, Jang SJ, Kim SH, Oh D, Kim MK, Kim SS, Kim M. Umbilical cord blood therapy potentiated with erythropoietin for children with cerebral palsy: A double-blind, randomized, placebo-controlled trial. Stem Cells. 2013;**31**(3):581–591

[29] Lee YH, et al. Safety and feasibility of countering neurological impairment by intravenous administration of autologous cord blood in cerebral palsy. Journal of Translational Medicine. 2012;**10**:58

[30] Purandare C, Shitole DG, Belle V, Kedari A, Bora N, Joshi M. Therapeutic potential of autologous stem cell transplantation for cerebral palsy. Case Reports in Transplantation. 2012;**2012**:825289

[31] Li M, Yu A, Zhang F, Dai GH, Cheng H, Wang X, An Y. Treatment of one case of cerebral palsy combined with posterior visual pathway injury using autologous bone marrow mesenchymal stem cells. Journal of Translational Medicine. 2012;**10**:100

[32] Luan Z, Liu W, Qu S, Du K, He S, Wang Z, Yang Y, Wang C, Gong X. Effects of neural progenitor cell transplantation in children with severe cerebral palsy. Cell Transplantation. 2012;**21**(Suppl 1):S91-S98

[33] Chen L, Huang H, Xi H, Xie Z, Liu R. Intracranial transplant of olfactory ensheathing cells in children and adolescents with cerebral palsy: A randomized controlled clinical trial. Cell Transplantation. 2010;**19**(2):185–191.

[34] Ramirez F, Steenblock D, Payne AG, Darnall L. Umbilical cord stem cell therapy for cerebral palsy. Medical Hypotheses and Research. 2006;**3**:679–686

[35] Payne G. Beneficial effects of subcutaneously injected human umbilical cord stem cells on cerebral palsy and traumatic brain injury in children and a posited mechanism. Medical Hypotheses and Research. 2005;**2**(3):497–501

[36] Sharma A, Sane H, Paranjape A, Gokulchandran N, Kulkarni P, Nagrajan A, Badhe P. Positron emission tomography–computer tomography scan used as a monitoring tool following cellular therapy in cerebral palsy and mental retardation – A case report. Case Reports in Neurological Medicine. 2013;**2013**:6 pages. Article ID 141983

[37] Sharma A, Kulkarni P, Sane H, Gokulchandran N, Badhe P, Lohia M, Mishra P. Positron emission tomography–computed tomography scan captures the effects of cellular therapy in a case of cerebral palsy. Journal of Clinical Case Reports. 2012;**2**:195

[38] Papadopoulos KI, Low SS, Aw TC, Chantarojanasiri T. Safety and feasibility of autologous umbilical cord blood transfusion in 2 toddlers with cerebral palsy and the role of low dose granulocyte-colony stimulating factor injections. Restorative Neurolology and Neuroscience. 2011;**29**(1):17–22

[39] Sharma A, Sane H, Kulkarni P, D'sa M, Gokulchandran N, Badhe P. Improved quality of life in a case of cerebral palsy after bone marrow mononuclear cell transplantation. Cell Journal. 2015;**17**(2):389–394

[40] Jensen A, Hamelmann E. First autologous cell therapy of cerebral palsy caused by hypoxic-ischemic brain damage in a child after cardiac arrest-individual treatment with cord blood. Case Report of Transplantation. 2013;**2013**:951827

[41] Wang L, Ji H, Zhou J, Xie J, Zhong Z, et al. Therapeutic potential of umbilical cord mesenchymal stromal cells transplantation for cerebral palsy: A case report. Case Report of Transplantation. 2013;**2013**:146347.

[42] Luan Z, Qu SQ, Liu WP, et al. Treatment of heteroptics after cerebral palsy with transplantation of human neural stem cells into cerebral ventricle in infants: 7 case report. Zhongguo Kangfu Lilun yu Shijian. 2007;**13**(12):1103–1105

[43] Wang X, Cheng H, Hua R, Yang J, Dai G, Zhang Z, Wang R, Qin C, An Y. Effects of bone marrow mesenchymal stromal cells on gross motor function measure scores of children with cerebral palsy: A preliminary clinical study. Cytotherapy. 2013;**15**(12):1549–1562

[44] Wang X, Hu H, Hua R, Yang J, Zheng P, Niu X, Cheng H, Dai G, Liu X, Zhang Z, An Y. Effect of umbilical cord mesenchymal stromal cells on motor functions of identical twins with cerebral palsy: pilot study on the correlation of efficacy and hereditary factors. Cytotherapy. 2015;**17**(2):224-31.

[45] Zali A, Arab L, Ashrafi F, Mardpour S, Niknejhadi M, et al. Intrathecal injection of CD133-positive enriched bone marrow progenitor cells in children with cerebral palsy: Feasibility and safety. Cytotherapy. 2015;**17**(2):232–241

[46] Mancías-Guerra C, Marroquín-Escamilla AR, González-Llano O, Villarreal-Martínez L, Jaime-Pérez JC, et al. Safety and tolerability of intrathecal delivery of autologous bone marrow nucleated cells in children with cerebral palsy: An open-label phase I trial. Cytotherapy. 2014;**16**(6):810–820

[47] Romanov YA, Tarakanov OP, Radaev SM, Dugina TN, Ryaskina SS, et al. Human allogeneic AB0/Rh-identical umbilical cord blood cells in the treatment of juvenile patients with cerebral palsy. Cytotherapy 2015;**17**(7):969–978

[48] Zhang C, Huang L, Gu J, Zhou X. Therapy for cerebral palsy by human umbilical cord blood mesenchymal stem cells transplantation combined with basic rehabilitation treatment: a case report. Global Pediatric Health. 2015;**2**:1–7

[49] Wang X, Hu H, Hua R, Yang J, Zheng P, et al. Effect of umbilical cord mesenchymal stromal cells on motor functions of identical twins with cerebral palsy: Pilot study on the correlation of efficacy and hereditary factors. Cytotherapy. 2015;**17**(2):224–231

[50] Shroff G, Gupta A, Barthakur JK. Therapeutic potential of human embryonic stem cell transplantation in patients with cerebral palsy. Journal of Translational Medicine. 2014;**12**(1):1

[51] Abi Chahine NH, Wehbe TW, Hilal RA, Zoghbi VV, Melki AE, Bou Habib EB. Treatment of cerebral palsy with stem cells: A report of 17 cases. International Journal of Stem Cells. 2016;**9**(1):90–95

[52] Fatemi SH, Aldinger KA, Ashwood P, Bauman ML, Blaha CD, et al. Consensus paper: Pathological role of the cerebellum in autism. Cerebellum (London, England) 2012;**11**(3):777–807

[53] Sajdel-Sulkowska EM, Xu M, McGinnis W, Koibuchi N. Brain region-specific changes in oxidative stress and neurotrophin levels in autism spectrum disorders (ASD). Cerebellum. 2011;**10**(1):43–48

[54] Wada N, Gronthos S, Bartold PM. Immunomodulatory effects of stem cells. Periodontology 2000. 2013;**63**(1):198–216

[55] Oswald J, Boxberger S, Jørgensen B, Feldmann S, Ehninger G, Bornhäuser M, Werner C. Mesenchymal stem cells can be differentiated into endothelial cells in vitro. Stem Cells. 2004;**22**(3):377–384

[56] Siniscalco D, Sapone A, Cirillo A, Giordano C, Maione S, Antonucci N, Autism Spectrum Disorders: Is Mesenchymal Stem Cell Personalized Therapy the Future? Journal of Biomedicine and Biotechnology, vol. 2012, Article ID 480289, 6 pages, 2012.

[57] Segal-Gavish H, Karvat G, Barak N, Barzilay R, Ganz J, Edry L, Aharony I, Offen D, Kimchi T. Mesenchymal stem cell transplantation promotes neurogenesis and ameliorates autism related behaviors in BTBR mice. Autism Research. 2016;**9**(1):17–32

[58] Sharma A, Gokulchandran N, Sane H, Nagrajan A, Paranjape A, et al. Autologous bone marrow mononuclear cell therapy for autism – An open label proof of concept study. Stem Cells International. 2013;**2013**:13 pages. Article ID 623875.

[59] Lv YT, Zhang Y, Liu M, Qiuwaxi J, Ashwood P, et al. Transplantation of human cord blood mononuclear cells and umbilical cord-derived mesenchymal stem cells in autism. Journal of Translational Medicine. 2013;**11**:196

[60] Bradstreet JJ, Sych N, Antonucci N, Klunnik M, Ivankova O, Matyashchuk I, M Demchuk, Siniscalco D. Efficacy of fetal stem cell transplantation in autism spectrum disorders: An open-labeled pilot study. Cell Transplantation, 2014;**23**:105–112

[61] Sharma A, Gokulchandran N, Badhe P, Kulkarni P, Mishra P, Shetty A, Sane H. An improved case of autism as revealed by PET CT scan in patient transplanted with autologous bone marrow derived mononuclear cells. Journal of Stem Cell Research and Therapy. 2013;**3**:2

[62] Sharma A, Gokulchandran N, Shetty A, Sane H, Kulkarni P, Badhe P. Autologous bone marrow mononuclear cells may be explored as a novel. Potential therapeutic option for autism. Journal of Clinical Case Reports. 2013;**3**:7

[63] Sharma A, Gokulchandran N, Sane H, Kulkarni P, Thomas N, Paranjape A, Badhe P. Intrathecal autologous bone marrow mononuclear cell transplantation in a case of adult autism. Autism Open Access. 2013;**3**:2

[64] Sharma A, Gokulchandran N, Sane H, Bhovad P, Biju H, Shetty A, Kali M, Badhe P. Cell therapy effects portrayed on positron emission tomography computerized tomography scan of the brain serve as a new dimension for autism: A case report. Journal of Paediatric Neurology. 2014;**12**:3

[65] Sharma A, Gokulchandran N, Shetty A, Kulkarni P, Sane H, Badhe P. Neuropsychiatric disorder tackled by innovative cell therapy – A case report in autism. Journal of Stem Cell Research Transplantation. 2014;**1**(1):4

[66] Sharma A, Gokulchandran N, Sane H, Patil A, Shetty A, Biju H, Kulkarni P, Badhe P. Amelioration of autism by autologous bone marrow mononuclear cells and neurorehabilitation: A case report. American Journal of Medical Case Reports. 2015;**3**(10):304–309

[67] Sharma A, Sane H, Gokulchandran N, Badhe P, Patil A, Kulkarni P, Paranjape A. PET-CT scan shows decreased severity of autism after autologous cellular therapy: A case report. Autism Open Access. 2016;**6**:169

[68] Sharma A, Chopra G, Gokulchandran N, Lohia M, Kulkarni P. Autologous bone derived mononuclear transplantation in Rett syndrome. Journal of Paediatric Practice. 2011; **15**(1):22–24

[69] Kang H, Schuman EM. Long-lasting neurotrophin-induced enhancement of synaptic transmission in the adult hippocampus. Science. 1995;**267**:1658–1662

[70] Sharma A, Sane H, Paranjape A, Gokulchandran N, Kulkarni P, Nagrajan A, Badhe P. Positron emission tomography–computer tomography scan used as a monitoring tool following cellular therapy in cerebral palsy and mental retardation – A case report. Case Reports in Neurological Medicine. 2013;**2013**:6 pages. Article ID 141983

[71] Wallace GQ, McNally EM. Mechanisms of muscle degeneration, regeneration, and repair in the muscular dystrophies. Annual Review of Physiology. 2009;**71**:37–57

[72] Meregalli M, Farini A, Colleoni F, Cassinelli L, Torrente Y. The role of stem cells in muscular dystrophies. Current Gene Therapy. 2012;**12**(3):192–205

[73] Torrente Y, Belicchi M, Sampaolesi M, Pisati F, Meregalli M, D'Antona G, Tonlorenzi R, Porretti L, Gavina M, Mamchaoui K, Pellegrino MA. Human circulating AC133+ stem cells restore dystrophin expression and ameliorate function in dystrophic skeletal muscle. The Journal of Clinical Investigation. 2004;**114**(2):182–195.

[74] Goetzl L, Darbinian N, Goetzl EJ. Novel window on early human neurodevelopment via fetal exosomes in maternal blood. Annals of Clinical and Translational Neurology. 2016;**3**(5):381–385

[75] Torrente Y, Belicchi M, Marchesi C, D'Antona G, Cogiamanian F, et al. Autologous transplantation of muscle-derived $CD133^{+}$ stem cells in Duchenne muscle patients. Cell Transplantation. 2007;**16**(6):563–577.

[76] Yang XF, Xu YF, Zhang YB, Wang HM, Lü NW, et al. Functional improvement of patients with progressive muscular dystrophy by bone marrow and umbilical cord blood mesenchymal stem cell transplantations. Zhonghua Yi Xue Za Zhi. 2009;**89**(36):2552–2556

[77] Mendell JR, Kissel JT, Amato AA, King W, Signore L, Prior TW, Sahenk Z, Benson S, McAndrew PE, Rice R, Nagaraja H. Myoblast transfer in the treatment of Duchenne's muscular dystrophy. New England Journal of Medicine. 1995;**333**(13):832–838

[78] Sharma A, Sane H, Badhe P, Gokulchandran N, Kulkarni P, Lohiya M, Biju H, Jacob VC. A clinical study shows safety and efficacy of autologous bone marrow mononuclear cell therapy to improve quality of life in muscular dystrophy patients. Cell Transplantation. 2013;**22**(Suppl 1):S127-S138

[79] Rajput BS, Chakrabarti SK, Dongare VS, Ramirez CM, Deb KD. Human umbilical cord mesenchymal stem cells in the treatment of Duchenne muscular dystrophy: Safety and feasibility study in India. Journal of Stem Cells. 2015;**10**(2):141

[80] Sharma A, Sane H, Gokulchandran N, Gandhi S, Bhovad P, Khopkar D, Paranjape A, Bhagwanani K, Badhe P. The role of cell transplantation in modifying the course of limb girdle muscular dystrophy: A longitudinal 5-year study. Degenerative Neurological and Neuromuscular Disease. 2015;**5**:93–102

[81] Skuk D, Goulet M, Roy B, Piette V, Côté CH, Chapdelaine P, Hogrel JY, Paradis M, Bouchard JP, Sylvain M, Lachance JG. First test of a "high-density injection" protocol for myogenic cell transplantation throughout large volumes of muscles in a Duchenne muscular dystrophy patient: eighteen months follow-up. Neuromuscular Disorders. 2007;**17**(1):38–46.

[82] Sharma A, Sane H, Paranjape A, Bhagawanani K, Gokulchandran N, Badhe P. Autologous bone marrow mononuclear cell transplantation in Duchenne muscular dystrophy – A case report. The American Journal of Case Reports. 2014;**15**:128.

[83] Sharma A, Paranjape A, Sane H, Bhagawanani K, Gokulchandran N, Badhe P. Cellular transplantation alters the disease progression in Becker's muscular dystrophy. Case reports in transplantation. 2013;**2013**:7 pages. Article ID 909328

[84] Sharma A, Sane H, Paranjape A, Badhe P, Gokulchandran N, Jacob V. Effect of cellular therapy seen on musculoskeletal magnetic resonance imaging in a case of Becker's muscular dystrophy. Journal of Case Reports. 2013;**3**(2):440–447

[85] Sharma A, Kulkarni P, Chopra G, Gokulchandran N, Lohia M, Badhe P. Autologous bone marrow-derived mononuclear cell transplantation in Duchenne muscular dystrophy. Indian Journal of Clinical Practice. 2012;**23**(3):169–172

[86] Sharma A, Sane H, Gokulchandra N, Sharan R, Paranjape A, Kulkarni P, Yadav J, Badhe P. Effect of cellular therapy in progression of Becker's muscular dystrophy: A case study. European Journal of Translational Myology. 2016;**26**(1):5522

[87] Sharma A, Sane H, Kaur J, Gokulchandran N, Paranjape A, Yadav J, Badhe P. Autologous bone marrow mononuclear cell transplantation improves function in a case of Becker's muscular dystrophy. American Based Research Journal. 2016;**5**(2):1–12

[88] Kang PB, Lidov HG, White AJ, Mitchell M, Balasubramanian A, et al. Inefficient dystrophin expression after cord blood transplantation in Duchenne muscular dystrophy. Muscle & Nerve. 2010;**41**(6):746–750

[89] Skuk D, Roy B, Goulet M, Chapdelaine P, Bouchard JP, Roy R, Dugré FJ, Lachance JG, Deschênes L, Senay H, Sylvain M. Dystrophin expression in myofibers of Duchenne muscular dystrophy patients following intramuscular injections of normal myogenic cells. Molecular Therapy. 2004;**9**(3):475–482

[90] Zhang C, Chen W, Xiao LL, Tan EX, Luo SK, Zheng D, Ye X, Li Z, Lu XL, Liu Y. Allogeneic umbilical cord blood stem cell transplantation in Duchenne muscular dystrophy. Zhonghua Yi Xue Za Zhi. 2005;**85**(8):522–525

[91] Hentze H, Graichen R, Colman A. Cell therapy and the safety of embryonic stem cell-derived grafts. Trends in Biotechnology. 2007;**25**(1):24–32

[92] Montoro J, Piñana JL, Moscardó F, Sanz J. Infectious complications after umbilical cord-blood transplantation from unrelated donors. Mediterranean Journal of Hematology and Infectious Diseases. 2016;**8**(1):e2016051

[93] Sharma A, Sane H, Paranjape A, Gokulchandran N, Takle M, et al. Seizures as an adverse event of cellular therapy in pediatric neurological disorders and its prevention. Journal of Neurological Disorders. 2014;**2**:164

[94] Minguell JJ, Pereira A, Bartholomew P, Lasala GP. The intrathecal infusion of mesenchymal stem cells into healthy rabbits is safe and devoid of neurological or clinical complications. Journal of Stem Cell Research and Therapy. 2011;**1**:104

[95] Xu Q, Zhang M, Liu J, Li W. Intrathecal transplantation of neural stem cells appears to alleviate neuropathic pain in rats through release of GDNF. Annals of Clinical and Laboratory Science. 2013;**43**(2):154–162

[96] Torrente. Y, Belicchi M, Marchesi C, D'Antona G. Autologous transplantation of muscle-derived CD133^{+} stem cells in Duchenne muscle patients. Cell Transplantation. 2007;**16**(6):563–577

[97] Fischer UM, Harting MT, Jimenez F, Monzon-Posadas WO, Xue H, Savitz SI, Laine GA, Cox CS Jr. Pulmonary passage is a major obstacle for intravenous stem cell delivery: The pulmonary first-pass effect. Stem Cells and Development. 2009;**18**(5):683–692

[98] Boltze J, Arnold A, Walczak P, Jolkkonen J, Cui L, Wagner DC. The dark side of the force – constraints and complications of cell therapies for stroke. Frontiers of Neurology. 2015; **6**:155

[99] Wang X, Zhang M, Feng R, Li WB, Ren SQ, Zhang J, Zhang F. Physical exercise training and neurovascular unit in ischemic stroke. Neuroscience. 2014;**271**:99–107

[100] Wahl P, Brixius K, Bloch W. Exercise-induced stem cell activation and its implication for cardiovascular and skeletal muscle regeneration. Minimally Invasive Therapy & Allied Technologies. 2008;**17**(2):91–99

[101] Acton PD, Zhou R. Imaging reporter genes for cell tracking with PET and SPECT. Quarterly Journal of Nuclear Medicine and Molecular Imaging. 2005;**49**(4):349–360

[102] Berti V, Mosconi L, Pupi A. Brain: Normal variations and benign findings in FDG PET/CT imaging. PET Clinics. 2014;**9**(2):129–140

[103] Mercuri E, Jungbluth H, Muntoni F. Muscle imaging in clinical practice: diagnostic value of muscle magnetic resonance imaging in inherited neuromuscular disorders. Current Opinion of Neurology. 2005;**18**:526–537

[104] Klotzenburg M, Yousry T. Magnetic resonance imaging of skeletal muscle. Current Opinion of Neurology. 2007;**20**:595–599

Chapter 9

Skeletal Muscle Dysfunction in Critical Illness

Yuki Iida and Kunihiro Sakuma

Additional information is available at the end of the chapter

http://dx.doi.org/10.5772/intechopen.69051

Abstract

Despite improvements in critical illness survival rates with recent developments in medical care, many patients still have long-term physical disabilities following stays in the intensive care unit (ICU). Critical illness–induced muscle weakness, so-called ICU-acquired weakness (ICU-AW), is a common occurrence in approximately 50% of critically ill patients in the ICU requiring mechanical ventilation for >7 days. ICU-AW contributes to increases in duration of mechanical ventilation and lengths of ICU and hospital stays and may persist among survivors for several years after discharge. Risk factors for ICU-AW include systemic inflammatory responses, severe sepsis, muscle inactivity, hyperglycemia, and use of neuromuscular blockers. Thus, the development of muscle wasting is suggested to be associated with pathophysiological alterations leading to an imbalance between muscle proteolysis and proteosynthesis through several cellular signaling networks. This chapter presents a review of the literature regarding critical illness–induced muscle wasting and describes potential treatment of excessive muscle catabolism.

Keywords: critical illness, ICU-AW, muscle proteolysis, immobilization, neuromuscular electric stimulation therapy

1. Introduction

Skeletal muscle is one of the most dynamic and plastic tissues and is the largest organ in the body. Skeletal muscle is primarily involved in mechanical activity, which depends on muscle fiber contractions required for posture, physical activity, and respiratory movement. However, skeletal muscle is not only a component of the movement system, but it accounts for approximately 40% of the total body weight and contains 50–75% of all body proteins. In general, muscle mass depends on the balance between muscle protein synthesis and degradation. Both processes are sensitive to a number of factors, such as nutritional status, hormonal balance, physical activity, and inflammatory stimulation. The loss of lean mass is accompanied by a concomitant loss of muscular strength, which can lead to physical disability and loss of function. Muscle mass, strength, and function are also correlated with each other, and their decreases can lead to poor health outcomes and mortality [1].

Although the rates of survival from critical illness have improved due to recent developments in medical care, many patients have long-term physical disabilities following a stay in the intensive care unit (ICU) [2]. Many survivors of critical illness complain of generalized muscle weakness, so-called intensive care unit acquired weakness (ICUAW), for several months to years after discharge from hospital [3] and have persistent exercise limitations [4–6]. Severe neuromuscular weakness and muscle wasting often complicate recovery from critical illness.

This primary myopathy can take various morphological forms, and is assumed to be triggered by both sepsis and other factors, including the extensive use of neuromuscular blocking agents and corticosteroids [7]. However, diagnosis of ICUAW is difficult in the ICU, because either the preexisting disorder or complications arising during the ICU stay can cause muscle weakness. In addition, the patient's aggravated condition when first admitted to the ICU precludes careful clinical examination of ICUAW. Consequently, the attention of the physician is directed toward survival, and can delay diagnosis of ICUAW.

Electrophysiological investigations of peripheral nerves and muscles can help in the diagnosis of ICUAW at an early stage and to define prognosis, but they are time consuming and require the availability of skilled personnel.

Therefore, a guided approach to diagnosis is valuable. Management of ICUAW rests on supportive treatment, treatment of ongoing sepsis and multiorgan failure, and control of hyperglycemia. Recent evidence indicates that early rehabilitation can be carried out safely and effectively to maintain the physical function of ICU patients, requiring little patients sedation [3].

This chapter describes the incidence rates, major risk factors, epidemiology, and electrophysiological histological features of ICUAW. Major advances in early rehabilitation and protocols with little or no sedation in the ICU, which improve the functional independence of patients, are also discussed.

2. Epidemiology

Physicians have become aware of the existence of long-term functional impairment after ICU treatment. Herridge et al. [5] examined loss of muscle mass, muscle weakness, and easy fatigue 1 year after recovery from acute respiratory distress syndrome (ARDS). Even among relatively young patients, half were unable to work for 1 year after discharge, and some dysfunction persisted even after 5 years [6]. Such disorders have been noted as ICUAW, and awareness regarding these issues has increased in intensive care medicine [7, 8].

ICUAW is divided into three categories: critical illness myopathy (CIM), critical illness polyneuropathy (CIP), and critical illness neuromyopathy (CINM) that combines features of the former two conditions [9]. CIP is a pathological condition characterized by significant axonal degeneration in sensory nerves, with no abnormalities in conduction velocity on electrophysiological analysis. In addition, the amplitudes of action potentials in muscle and sensory nerves are reduced [10]. Peripheral nerve microangiopathy due to inflammatory responses is involved in the pathophysiology of CIP [11, 12]. CIM is characterized by general muscle weakness and residual sensory function [13]. Catabolism accompanied by skeletal muscle disruption is often observed as the mechanism underlying its development. The muscle disorder caused by systemic inflammation and inactivity due to bed rest would be major factors [14, 15]. However, diagnosis of CIP and CIM is difficult because either existing disorders or complications occurring during the ICU stay causes muscle weakness. In addition, patient survival takes precedence in the ICU, and diagnosis of CIP and CIM is delayed or overlooked. Therefore, it is difficult to accurately determine the incidence rates of CIP and CIM [12, 16, 17]. Intriguingly, recent data suggest that CIP and CIM frequently coexist, a condition that has been termed CINM [18]. The pathophysiology of CINM/ICUAW is complex and includes the sequelae of bed rest, the effects of critical illness–induced cytokine production, and possibly the interplay of drugs, such as neuromuscular blocking agents and corticosteroids (**Figure 1**). Protein-energy malnutrition, electrolyte imbalances, and glutamine deficiency also play roles in the critically ill.

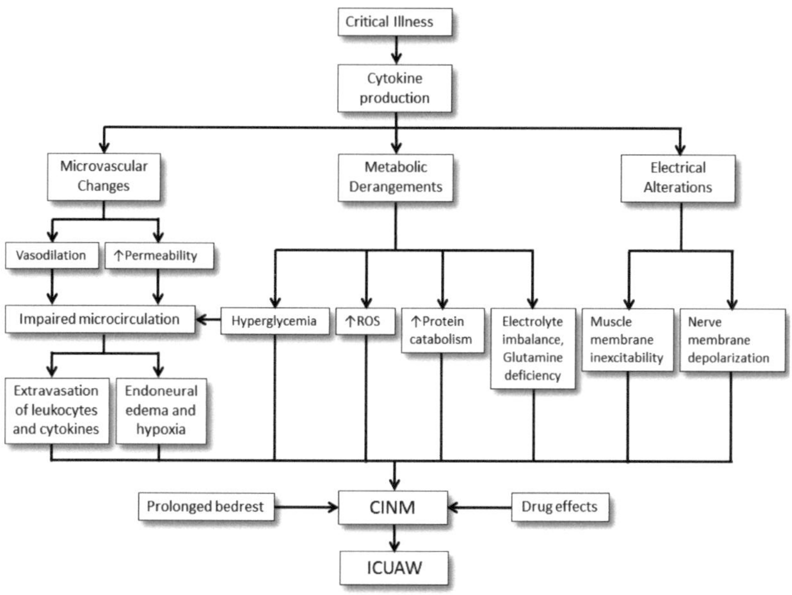

Figure 1. Pathophysiology of CINM/ICUAW. CINM, critical illness neuromyopathy; ICU, intensive care unit; ICUAW, ICU-acquired weakness; NMBA, neuromuscular blocking agent; ROS, reactive oxygen species.

Current estimates indicate that 70–80% of critically ill patients develop CIP, with a comparable percentage presumably developing CIM [16, 19, 20]. In subpopulations in which sepsis is complicated by multiple organ failure (MOF), the incidence rates of CIP and CIM could even reach 100% [20]. Approximately two-thirds of patients with ARDS exhibit these neuromuscular disorders [21], and in unselected patients that have required mechanical ventilation for at least 4 days, the incidence rates of CIP and CIM range from 25 to 33% on clinical evaluation, and can reach 58% on electrophysiological evaluation [4, 17]. Furthermore, 49–77% of patients will have ICUAW when treated in an ICU for ≥7 days [22, 23].

ICUAW leads to prolonged physical disabilities that persist for months or years after ICU discharge [18, 24–26]. Nearly one-third of patients with ICUAW have difficulty in recovering independent walking or spontaneous breathing [27]. The effects of critical illness and ICU stay can result in serious weakness [3]. In the CRIMYNE study, the findings from the 1-year follow-up cohort study showed that patients with CIM recovered within 6 months, whereas those with CIP had a slower recovery or did not recover [17]. Mortality is increased in patients with CIP [4, 28].

The incidences of CIP and CIM are not clear, because of the wide variation in patient populations, risk factors, and diagnostic criteria, as well as in the timing of assessment [29]. In patients with mechanical ventilation beyond 7 days and with the presence of systemic inflammatory response syndrome (SIRS), the incidences of CIP and CIM were 33% on clinical assessment [22] and 30–58% on electrophysiological assessment [4, 10, 21]. The incidence rates were 34–60% in patients with ARDS [21], 24–77% in those with ICU stay beyond 1 week [10, 21, 22], 56–80% in those with multiorgan failure [15], and 100% in those with septic shock or severe sepsis [20]. A systematic review showed evidence of CIP and CIM in 46% of adult ICU patients that had lengthy periods of mechanical ventilation, sepsis, or multiorgan failure [10].

3. Risk factors for ICUAW

The ICUAW is regulated by several risk factors such as multiple organ failure, inactivity, hyperglycemia, corticosteroids, and neuromuscular blocking agents, although the pathogenic mechanism of ICUAW remains unclear (**Table 1**) [7, 30].

In particular, multiple organ dysfunction and immobility are important factors causing muscle dysfunction in patients in critical condition.

Multiple organ failure (MOF) is the major risk factor for ICUAW, but one needs to interpret it carefully. Neuromuscular organ failure of sepsis is the central element of the pathological condition in ICUAW. Indeed, sepsis, SIRS, and MOF develops the CIP [31]. Other diseases caused by SIRS, such as sepsis, induce mitochondrial injury [32], sodium channel inactivity [33], and have adverse effects on neuromuscular activity. Skeletal muscle proteolysis is the main pathogenic mechanism of CIM [28, 34]. The specific pathology of ICUAW reflects the structural deterioration of peripheral nerves and skeletal muscles, and not only disuse muscle atrophy [35].

Immobility also facilitates muscle wasting in critical illness. The conditions of low mechanical load, such as bed rest, immobilization, and disuse, induce marked losses of skeletal muscle mass, strength, and physiological function. Disuse muscle wasting involves a complicated cytokine and inflammatory response. Even a short period of bed rest for 5 days will results

SIRS/sepsis
Multiorgan failure
Hyperglycemia
Renal replacement therapy
Catecholamine administration
Female sex
Duration of mechanical ventilation
Corticosteroids
Neuromuscular blocking agents

SIRS, systemic inflammatory response syndrome.

Table 1. Risk factors implicated in the development of ICU-acquired weakness.

in significant decreases in the size of muscle fibers (3.5–10%) and muscle strength (9–13%) [36, 37]. However, as patients with ICUAW are usually under sedation, it is not clear how much sedation-induced inactivity affects the condition, apart from the severity of the disease. Therefore, to elucidate the pathology of ICUAW, factors of inactivity should be examined separately. Griffiths et al. [38] performed sustained passive exercise on one lower limb in a patient with severe respiratory failure in neuromuscular block, and examined the effects of intervention using the contralateral lower limb as a control limb. The results indicated that muscle strength and muscle protein level were significantly higher in the intervention limb. Thus, ICU patients suffer from long-term inactivity, and the methodology of physical activity in ICU should be reconsidered.

Diaphragmatic weakness, pulmonary injury, and atrophy develop rapidly during mechanical ventilation with sedation [39]. The duration of mechanical ventilation is independently associated with severe limb weakness or electrophysiological evidence of CIP [33, 40]. Respiratory muscle weakness also extends mechanical ventilation duration in the ICU. Respiratory muscle weakness elicits infection or sepsis, disease severity, and peripheral weakness [41, 42]. These data also suggest that ICUAW includes respiratory muscle weakness, because the diaphragmatic nerve and diaphragm also exhibit electrophysiological abnormalities similar to peripheral nerves and muscles [43].

Neuromuscular blocking agents and corticosteroids dose similar risks for ICUAW [44]. The risk of acute myopathy by the treatment with corticosteroid and neuromuscular blocking agent appears to increase after 24–48 hours of therapy. In addition, corticosteroids are the main determinants of impaired ability to exercise at 3 months in patients with critical illness myopathy [5]. After the first case report of profound frailty in asthma patients treated with corticosteroids and neuromuscular blocking agents [45], many similar findings were reported indicating that these two substances [28, 46] contribute to CIP/CIM or ICUAW. In animal studies, denervated and steroid-treated animals showed muscle changes similar to those observed in critical patients [47]. However, some studies have failed to confirm the adverse effects of corticosteroids [22, 28, 48] or neuromuscular blocking agents [49]. The interaction between these drugs and ICUAW would ascribe to use as well as to complicated relationships with dose, timing, and other risk factors for ICUAW.

The other independent risk factors for ICUAW identified in previous studies are disease severity [22, 50], use of vasopressors and catecholamine [40], length of stay in the ICU [34, 40], renal failure and renal replacement therapy [50], hyperosmolality [28], parenteral nutrition [28], low serum albumin level [34], and neurological disorders [34]. Hyperglycemia was shown to be an independent risk factor for the electrophysiological and clinical signs of ICUAW, and increased insulin dose reduced the incidence of ICUAW [51]. Several studies indirectly demonstrated that the severity of disease is correlated with ICUAW, as reflected by acute inflammatory mediators and the use of vasopressors [40, 52]. Age is also an important risk factor for ICUAW [51]. As hospitalization in the ICU is often unpredictable, it is difficult to confirm the state of muscle function before onset. The motor function before hospitalization depends on many factors, but the muscle strength and muscle mass before onset reflect age to some extent.

4. Critical illness–induced muscle wasting

There are at least two phases, early and late phase, in the pathophysiological process of myopathy occurring in critically ill patients (**Figure 2**). The early phase progresses rapidly between the 3rd and 5th day after onset, whereas the late phase is slowly sustained from 5 days after onset. The following section describes how sepsis and inactivity, which are the main factors of ICUAW, influence myopathy in each phase.

4.1. Early phase

Pathophysiologically, the initial response to severe infection is in the activation of antigen-presenting immune cells. This occurs via pattern recognition receptors [53–55] with other immune mechanisms, such as cytokine release, endothelial and complement activation, and release of oxygen radicals. Inflammatory cytokines [tumor necrosis factor alpha (TNF-α), interleukin (IL)-6, and IL-1β] and complement factors (C3a and C5a) act as important mediators at this stage of inflammation [54]. Inflammatory stimulation causes muscle wasting mainly by degradation of muscle proteins [56]. Muscle proteins break down into amino acids and some are used in the liver for the synthesis of glutathione and acute proteins. Muscle proteins to be degraded are derived from myofibrillar proteins (actin, myosin), which account for 60–70% of muscle proteins [56]. Therefore, stressed patients lose 250 g/day of muscle protein corresponding to a muscle mass between 750 and 1000 g [57].

The muscle proteolysis in the early phase is caused by several cell signaling systems, but the major pathway of proteolysis activated in animal models of inflammatory muscle wasting are the calpain system and ubiquitin-proteasome system (UPS) (**Figure 3**). Calpain, a calcium-dependent protease, is activated first [58]. In septic rats, calpain activity seems to be increased by 70% as measured by the degradation products of azocasein in muscle extracts [59]. Although inhibition

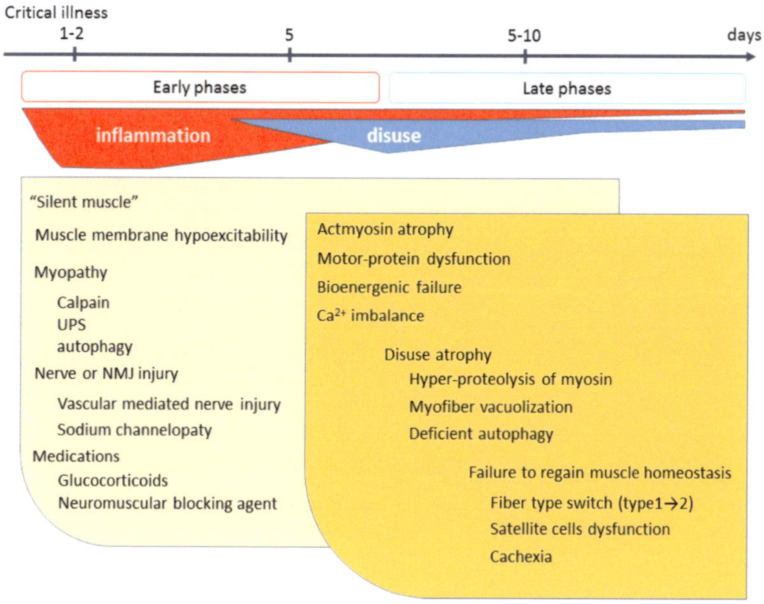

Figure 2. Mediators of critical illness-induced muscle dysfunction. Skeletal muscle atrophy is the most universal feature of the early phase, which is driven fundamentally by inflammation and disuse. Other factors, such as neuropathic injury and medications, can exacerbate atrophy and independently cause muscle dysfunction. Therefore, inhibiting muscle protein degradation is the most promising potential early-phase therapy. The late phase is marked by cessation of inflammation-induced muscle proteolysis, and therefore potential treatments at this time point will differ. Mediators of the late phase would involve persistence of some early-phase injuries or a failure to regain muscle homeostasis following the early phase. Late-phase dysfunction may be compounded by underlying premorbid neuromuscular defects. NMJ, neuromuscular junction; UPS, ubiquitin-proteasome system.

of calpain function cannot suppress muscle weight reduction, the sarcomere structure and contractile tension per unit area of muscle fibers are maintained at normal levels [60]. These findings suggest that the calpain system targets cytoskeletal proteins that maintain the sarcomere structure. Activated calpain does not directly degrade contractile proteins, such as actin and myosin, but degrades cytoskeletal proteins, such as titin and nebulin [61]. The activity of calpastatin, a calpain system inhibitor, is also decreased by 40–60% in the muscles of septic rats [62]. As a result, myofibrils, such as actin and myosin, are released into the cytoplasm in the monomeric state.

At the next stage, activation of the UPS rapidly degrades the myofibrils. UPS is the major decomposition system in various myopathies causing muscle atrophy, such as inactivity, inflammation, cell energy stress, and malnutrition. The forkhead box O (FOXO) family of transcription factors stimulates the expression of two important regulators of UPS-mediated proteolysis of ubiquitin ligase atrogin-1 and muscle-specific RING finger protein-1 (MuRF1) [63–65]. In animal models, both MuRF1 and atrogin-1 recruitment, and UPS activation, mainly contribute to the loss of muscle mass, such as inactivation-induced atrophy, acute illness, chronic disease, and CIM [63, 66, 67]. In human studies, mRNA expression levels of components of the UPS are increased in septic muscle. Muscle-specific E3 protein, MuRF1, and atrogin-1 in the vastus lateralis of critical patients are elevated at both mRNA and protein

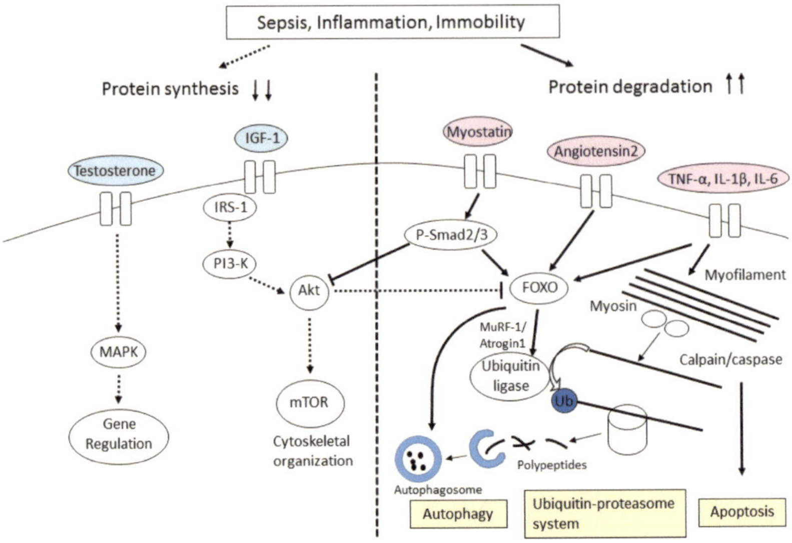

Figure 3. Muscle protein homeostasis in critical illness. The diagram depicts changes in protein metabolism caused by pathophysiological processes in critical illness, including sepsis, inflammation, and immobility. Upper circles denote altered regulation of major steps in protein synthesis (left) and degradation (right). PI3-K, phosphatidylinositol-4,5-bisphosphate 3-kinase; Akt, also known as protein kinase B; mTOR, mechanistic target of rapamycin; MAPK, mitogen-activated protein kinase; FOXO, forkhead box-O; Ub, ubiquitin; IRS, insulin receptor substrate.

levels [68]. The chymotrypsin-like peptidase activity of the membrane-associated proteasome is increased by 30% in limb muscle of critical patients [69].

Similar mechanisms contribute to respiratory muscle weakness in the acute phase. MuRF1 activity in the diaphragm is increased in critical patients under artificial respiratory management [70]. Postoperative respiratory muscle weakness would be caused by increased postoperative inflammatory cytokine production [71]. Mechanical ventilation as a substitute for diaphragm immobility is an important trigger, especially when combined with sedation, leading to weakness of the diaphragm [70].

The major etiology in the acute phase of CIP may be axonal dysfunction due to membrane depolarization disorders [72]. A disorder of the blood-nerve barrier is a pathogenic event induced in the absence of ultrastructural injury in the peripheral nerves [73]. Especially in peripheral nerves, the lack of autoregulation in capillary vessels leads to impairment of microcirculation. When the vascular permeability is increased, edema occurs in the endoneurium and ischemia develops in the tissue, resulting in microcirculatory insufficiency of peripheral nerve vessels [72]. In particular, damage to structural proteins of axons with high energy demand causes CIP with axonal degeneration of peripheral nerves.

4.2. Late phase

Risk factors of muscle wasting in severe patients directly or indirectly induce muscle inactivity. In the late phase, inactivity is the main cause of skeletal muscle contraction. Inactivity directly affects muscle wasting, even if there are no systemic inflammatory changes [74].

Some studies demonstrated the adverse effects of one-leg suspension and casting in bed rest on skeletal muscle mass in humans—the muscle mass and the cross-sectional muscle area were significantly decreased in the inactive leg. Muscle atrophy begins within hours of commencement of sleep and deep sedation. Muscle mass and muscular strength decline greatly within 10 days of rest, especially in the lower limbs, even in healthy people [75].

Anomalies in action potentials due to inactivity occur within several hours. As the general state of health recovers, potential anomalies readily improve [33]. However, this improvement reaches a plateau within 3–6 months, and some patients do not recover normal function even in 1–2 years. Little is known about the pathophysiology of CIP in this latter type of sustained ICUAW.

Electrical nonirritability of muscle has been reported in rat denervation and steroid administration models [76]. The primary cause of hypoexcitability of steroid-denervated fibers is the combination of resting potential depolarization and hyperpolarized shift in the voltage dependence of NaV1.4 sodium channel inactivation [77]. Depolarization of stationary membranes causes muscle inactivity and could be an important mechanism accounting for significant muscle weakness during immobility [78]. Upregulation of NaV1.5 sodium channels has been demonstrated in the muscle of rats with chronic sepsis, suggesting that several risk factors lead to muscle electrical nonexcitability [79]. A negative shift in sodium channel gating in peripheral nerves of muscle is distinguished from CIM as a characteristic disorder in CIP [80].

Recently, autophagy, a bulk degradation system of the cytoplasmic matrix, was shown to potently induce muscle atrophy [81]. Animal models of muscle atrophy (demyelination, sepsis, starvation) show increased autophagy, and FOXO transcriptional regulators modulate expression of several autophagy supporting genes [82–84]. However, autophagy dysfunction in mice caused enhanced oxidative stress inducing myofiber protein aggregates, abnormal muscle mitochondrial accumulation, and myofiber degeneration [85]. Thus, titration of the degree of autophagy appears to be essential for maintaining healthy muscle mass. That is, excessive autophagic activity leads to muscle atrophy, whereas toxic products accumulate and lead to muscle degeneration under conditions of autophagy failure.

Some studies showed UPS activation in acute CIM and septic human atrophic skeletal muscle [68, 86, 87], but others showed muscle protein degradation with low levels of MuRF1/atrogin-1 [88]. These discrepancies seem to be due to the dynamics of expression of UPS component mediators in muscle wasting. For example, in animal models, MuRF1/atrogin-1 mRNA levels increase only in short-term denervated or unweighted muscles, but do not increase over a long period of time [66]. Likewise, MuRF1 mRNA expression was also increased in the lateral vastus muscle of healthy volunteers with short-term rather than long-term periods of inactivity [89]. In chronic complete spinal cord injury patients, there were significant reductions in atrogin-1, MuRF1, and myostatin mRNA levels, and in FOXO1, FOXO3a, and atrogin-1 protein levels [90]. Therefore, another pathway seems to be promoted in the late phase instead of UPS.

Hussain et al. [91] reported that autophagy contributes to the induction of diaphragm muscle proteolysis in ICU patients. Induction of autophagy increased protein oxidation and enhanced expression of the FOXO1 gene, but not the FOXO3A gene. Controlled mechanical ventilation also triggered the inhibition of both Akt expression and FOXO1 phosphorylation. In addition, the decrease in mature autophagic vesicles and accumulation of p62 (a protein degraded by autophagy) indicated deficiency of autophagy in the muscle of critically ill patients [92]. In severe conditions, autophagy would be the cause of muscle atrophy, and participate in the mediation of acute or persistent CIM.

Nuclear import of FOXO activates both UPS and autophagy pathways, and promotes muscle protein degradation of myotube cells, 70% of which is due to autophagy [61]. The expression of autophagy-related genes in skeletal muscle is markedly increased by fasting and denervation, and autophagy appears to be an important degradation system for chronic muscle atrophy in sarcopenia and cachexia [93]. In addition, FOXO and p38 mitogen-activated protein kinase (MAPK) induce expression of the muscle-specific E3 gene as well as expression of the autophagy-related gene Atg7 [94]. Furthermore, autophagy-related molecules, such as Beclin-1 and LC3-II proteins, are elevated in skeletal muscle under denervation for 7 days [95]. The elevated levels of LC3 and Beclin-1 mRNA expression were also reported in skeletal muscle after 3 or 7 days of denervation [64]. Under such conditions of inactivity, such as the late phase, autophagy seems to modulate muscle atrophy accompanying inactivity [96].

Critically ill patients lose muscle as a result of an inability to maintain rates of protein synthesis above those of protein breakdown [97]. Decline in muscle protein synthesis is observed very early in the period of inactivity. Studies using inactivity models, such as rat tail suspension and casting fixation, suggest that muscle protein synthesis declines rapidly within 6–24 hours after commencement of inactivity and that level is maintained during inactivity [98]. Therefore, the decrease in muscle protein synthesis contributes to muscle atrophy in the very early stages of inactivity. Insulin-like growth factor-1 (IGF-1) secreted from skeletal muscle along with physical activity binds to the IGF-1 receptor and promotes the activation of PI3K-Akt-mammalian target-mediated signaling of the rapamycin (mTOR) pathway. Subsequently, protein synthesis is promoted through phosphorylation of p70S6K involved in the initiation of downstream translation and inactivation of the translational repressor factor 4E-BP1. The PI3K/Akt pathway is an important network known to induce muscle protein synthesis and muscle growth [60]. Interestingly, Akt activation has been shown to block the progression of FOXO nuclear translocation and atrophy, demonstrating mutual signaling between muscle atrophy and hypertrophy induction [99].

The levels of Akt1, mTOR, p70S6K, and 4E-BP1 phosphorylation in the soleus muscle of rats are decreased by tail suspension for 7 days [100]. Signaling pathways regulating muscle protein synthesis are repressed in critically ill patients. For the mTOR pathway, a sepsis model showed interference with protein translation and inhibition of protein synthesis [101]. Protein phosphorylation was decreased for all signaling proteins of Akt1, PKB, GSK3, mTOR, p70S6K, and 4E-BP1 in a study of vastus lateralis muscle biopsies in 10 critical patients [86]. Therefore, in inactive skeletal muscles in which mechanical stress is relieved, decreases in IGF-1/PI3K/Akt signals cause reductions in muscle protein synthesis. These findings indicate that the signal transduction activity promoting protein translation is reduced in inactive patients.

5. Interventions for ICUAW

Effective interventions for ICUAW have not yet been established. As a strategy, avoiding or diminishing iatrogenic risk is of primary importance [44]. Risk factors for the occurrence of ICUAW include sepsis, hyperglycemia, inactivity, malnutrition, corticosteroids, and the use of neuromuscular blockers. All of these are risk factors causing hypercatabolic states, and avoiding these factors is a useful intervention strategy for muscle dysfunction.

5.1. Intensive insulin treatment

Several Randomized controlled trials (RCTs) have been reported for glycemic control in the ICU [102–104]. Many large clinical trials indicated that hyperglycemia should be avoided [105]. Increases in insulin resistance causing hyperglycemia occur frequently in critically ill patients, and are more apparent in CIM patients [106]. As a mechanism, GLUT-4 translocation impairment to myocytes would decrease glucose supply in patients. In a study based on muscle biopsy specimens, intensive insulin therapy (IIT) improved insulin resistance and GLUT-4 translocation in skeletal muscle [107]. RCT was performed to evaluate the effects of IIT (glycemic target 80–110 mg/dl) on neuromuscular function in ICU patients. IIT reduced the incidence of severe CIP from 49 to 25% in electrophysiological screening studies of long-term hospitalized surgery patients [40]. In another study, ITT decreased the duration of mechanical ventilation [108]. The relative risk for developing CIP by IIT was 0.65 (95% confidence interval 0.55–0.77) [44]. Its effect was ascribed to glycemic control rather than insulin dose [40], and it was observed only in patients in whom blood glucose was controlled within the normal range [103]. Similar findings were obtained in a retrospective study on the incidence of electrophysiological abnormalities before and after IIT [109].

On the other hand, Derde et al. [110], using muscle biopsy samples, demonstrated that IIT does not affect myofiber size, myofibrillar protein synthesis ability, or muscle proteolytic markers. ITT is considered to primarily have a neuroprotective role [107], but no neurological data are available. In the NICE SUGAR trial, it was questionable whether the blood glucose level should be strictly controlled within the normal range as a general treatment in critical care patients. In addition, intervention of IIT in critically ill patients increases the incidence of hypoglycemia and mortality rate [102]. These results would be due to methodological differences among the trials [111]. A strategy to reduce hyperglycemia without the risk of hypoglycemia is appropriate to reduce the incidence of ICUAW.

5.2. Early mobilization

In both the early and late phase, bed rest and mechanical unloading induce catabolism, muscle atrophy, and weakness. Therefore, minimizing the duration of inactivity would reduce the incidence of ICUAW [112, 75].

The first step to shorten the duration of inactivity is to reduce sedation. Reducing sedation seems to have a number of beneficial effects, including shortened period of mechanical ventilation and ICU stay, and lessened delirium [113]. However, there have been no studies of its role in prevention of muscle function decline. Reduction of sedation could contribute to early mobilization in critically ill patients. The combination of approach to minimize sedation and early mobilization was investigated in critically ill patients with long-term mechanical ventilation [114]. Early mobilization and standard treatment were compared in 104 patients that received daily sedation interruption. Early mobilization improves exercise function at discharge, and shortens the duration of delirium and the duration of mechanical ventilation [114]. As there was no significant decrease in the incidence of ICUAW, reducing sedation would not improve the muscle strength itself but would help the patient to exercise efficiently.

To shorten the length of inactivity, early rehabilitation is carried out even with the use of a life-support device, such as mechanical ventilator or left ventricular assist device. There have been several reports that early rehabilitation is safe and feasible even in the early phase in the ICU, and is beneficial for reducing ICU and hospital stay [115, 116]. RCT comparing "cycling exercise 20 minutes at the bedside every day" and "standard physical therapy" during ICU stay did not report the occurrence of muscle weakness, but the isometric knee extensor strength and 6-minute walking distance were significantly higher in the intervention group [117].

However, there are several barriers to early rehabilitation, which may prevent the penetration of this approach [116, 118]. Although many ICUs have successfully implemented early rehabilitation into routine clinical care, widespread implementation remains low, with only 8–12% of mechanically ventilated patients mobilized out of bed as reported in two large multisite point prevalence studies [119, 120]. Although it is impossible to change illness severity and consciousness level of the patient, sedation levels [121] and intentions for treatment of therapists [122] can be corrected appropriately. For implementation of rehabilitation, appropriate protocols in the ICU are necessary to confirm the balance between beneficial effects and risk of mobilization, and to select the correct treatment intensity for the patients.

A study of 49 people that were not weaned from the ventilator for at least 14 days showed that whole-body rehabilitation and respiratory muscle training could improve muscular strength, ventilator weaning, and functional status [123]. Intriguingly, Connolly et al. [124] reported that rehabilitation after discharge is beneficial. Physical therapy during the recovery phase in ICUAW plays an important role, as the disuse syndrome during the late phase and anabolic resistance are added to the muscle dysfunction following early phase. As this problem has not been resolved, it is unclear how much such a strategy will promote recovery from ICUAW.

5.3. Neuromuscular electrical stimulation

Critically ill patients in the ICU have unstable cardiopulmonary dynamics and mental state, and in some cases are controlled by ventricular assist devices or hemodialysis. Therefore, not all patients undergo active physical activity. Under these restrictions, neuromuscular electrical stimulation (NMES), which involves muscle contraction by outside electrical stimulation, is used.

Some studies indicated the effects of NMES on patients in the ICU. In septic patients with mechanical ventilator management, NMES was applied to one lower limb and the other was used as a control limb. Muscle strength and muscle protein synthesis were improved more on the stimulated side limb, although the muscle cross-sectional area and muscle mass decreased in both limbs [125, 126]. A previous RCT using NMES indicated increased muscle strength and decreased mechanical ventilator duration and the incidence of ICUAW decreased from 39.3 to 12.5% in the NMES intervention group [127].

NMES intervention for Chronic obstructive pulmonary disease (COPD) patients under mechanical ventilation improved the muscle strength and reduced the number of days needed to transfer from bed to chair [128]. Another RCT in COPD patients within the ICU did not show the occurrence of ICUAW in patients treated with NMES [129]. In addition, in the NMES group, the quadriceps muscle strength measured with a dynamometer and the walking distance were increased [130]. Although not performed in patients in the ICU, NMES intervention was shown to increase muscle strength and exercise tolerance [131]. NMES for critically ill patients in the ICU would ameliorate muscle weakness and improve mobility [132, 133].

Several basic studies examined the direct effects of NMES on muscle function in the ICU. Factors of muscle dysfunction, such as ICUAW, include neuropathy and muscle protein degradation caused by inflammation, peripheral microcirculation disorder, increased insulin resistance, and abnormal energy metabolism due to mitochondrial dysfunction [134]. NMES improves these factors that promote muscle catabolism [135–139]. In addition, NMES for patients undergoing cardiac surgery reduced the postoperative excretion of 3-methylhistidine as an indicator of muscle proteolysis [140]. A preliminary study using muscle biopsy specimens from patients at risk for ICUAW indicated that NMES could improve AMPK activation, glucose utilization, and GLUT-4 translocation [106].

Taken together, these findings indicated that NMES can inhibit catabolism and promote the anabolic pathway (**Figure 4**). As NMES can stimulate muscle contraction quantitatively and independent of intention in the patient, performance of NMES from the early phase is the most effective intervention for ICUAW. However, NMES research data should be interpreted carefully because of significant differences in baseline characteristics among patients, including APACHE II score and certain comorbidities [141]. There are several methodological problems, such as small sample size and incomplete results, and therefore these findings should be confirmed in large-scale trials. Although the amount of stimulation is unclear, it is important to maintain muscle contraction even at the early phase of critical illness.

5.4. Nutritional strategies

Malnutrition develops rapidly in critically ill patients due to dysfunction of the gastrointestinal tract. Hence, rich nutrition management in the acute phase may be considered, but the scientific evidence is not yet clear.

Although patients undergoing esophagectomy or pancreaticoduodenectomy did not receive enteral nutrition during the 6 days postoperatively, there were no significant differences in grip strength, respiratory muscle strength, or recovery of walking ability, compared with nutritionally fed patients after jejunostomy [142]. In the EPaNIC study, 4640 patients were randomized to early parenteral replacement therapy (early PN group) or tolerating caloric deficiency for first week in ICU (late PN group) [143]. The late PN group

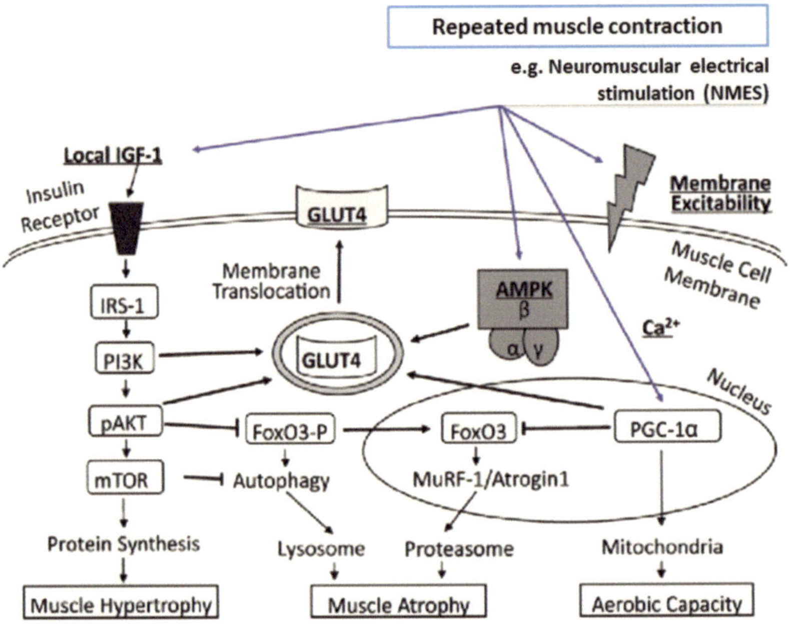

Figure 4. Suggested beneficial effects of neuromuscular electrical stimulation (NMES) with regard to muscle hypertrophy, atrophy, aerobic capacity, membrane excitability, and membrane translocation of GLUT4. NMES may preserve membrane excitability. Membrane translocation of GLUT4 is regulated by IGF-1, AMPK, PGC-1α, and its downstream targets, which may all be affected by NMES. Atrophy gene expression (MuRF-1, atrogin-1) increases upon dephosphorylating of FOXO3 transcription factors, which is inhibited by downstream insulin signaling.

showed promotion of recovery, decreased mechanical ventilator duration, and reduced complications compared to the early PN group. In addition, whole-body muscle strength assessment using the Medical research council scores was performed in patients predicted to be at high risk of developing ICUAW [144]. Evaluation of muscle strength was conducted three times a week from day 8 until ICU discharge or death. The incidence of ICUAW was significantly lower in the late PN group compared with the early PN group. In the muscle biopsy specimens of 122 patients with EPaNIC, LC3II and LC3I ratios associated with autophagosome formation were higher in patients in the late PN group than the early PN group [144]. This study indicated that the muscle fiber autophagy pathway was more efficient in the late PN group.

Another large RCT was performed in 1372 ICU patients to compare conventional management with early parenteral nutrition. This study indicated a greater degree of muscle wasting in patients under conventional management [145]. The Eden trial examined enteral nutritional management versus nutritional feeding in 1000 ICU patients with acute lung injury [146]. At 6- and 12-month follow-up, the physical abilities of 174 survivors were significantly lower than the predicted values. However, no significant differences were found in MRC total score, grip strength, maximum inspiratory pressure, 6-minute walking distance, or quality of life between the two groups [147]. Therefore, high energy intake from the early phase showed no evidence of improved physical function, but rather may worsen muscle wasting.

Treatment with various supplemental nutritional components has been studied. Glutamine concentrations in plasma and skeletal muscle are low in critical patients and are independent predictors of mortality [57]. Glutamine is an essential amino acid and its supplementation improves muscle function [57, 148]. A meta-analysis of glutamine supplementation in critically ill patients reported beneficial effects on prognosis [149]. In contrast, two RCTs indicated no such effect of glutamine supplementation [150, 151]. Although the effects on muscle function have not been studied, an increase was observed in mortality in critical patients with multiple organ failure [151]. The use of other micronutrients against oxidative stress is sensible from the pathophysiological point of view. Two meta-analyses concluded that antioxidant micronutrients may be beneficial for critically ill patients [152, 153]. Preliminary data from animal studies indicated that mitochondrial reactive oxygen species (ROS) is an important factor in the progression of diaphragm atrophy and contractile dysfunction occurring during mechanical ventilation [154]. However, the effects on muscle weakness were not examined, and a recent large RCT showed no beneficial effect of antioxidants in critical patients with multiple organ failure [151]. Therefore, early nutritional therapies enriched with immunomodulatory nutrients would not reduce the incidence of ICUAW compared with standard nutritional therapy for critically ill patients [155].

6. Summary and future directions

ICUAW is a common complication and an important contributor to the physical disabilities persisting in ICU survivors. Although various interventions have been used to prevent the adverse effects of ICUAW, no established therapy is available. Early rehabilitation can be an important preventative therapy for ICUAW. "Silent muscle" in the early phase of ICUAW leads to muscle catabolism through the protein degradation signaling pathway. Strategies aimed at minimizing the duration of immobilization would contribute to the suppression of muscle wasting during critical illness. Promotion of intermittent muscle contraction, such as NMES, would mitigate the severity and incidence of muscle wasting. Sakuma et al. [156] reported that resistance training combined with amino acid-containing dietary supplements would be the best way to prevent the muscle wasting and weakness, including sarcopenia. This proposal can also be applied to preventive strategies for ICUAW, especially in the late phase. Further research is required to design novel rehabilitation strategies for initiating anabolic reaction and improving muscle function in patients that cannot actively participate in physical therapy in the acute phase of critical illness.

Author details

Yuki Iida[1]* and Kunihiro Sakuma[2]

*Address all correspondence to: y2iida@na.commufa.jp

1 Department of Rehabilitation Medicine, Kainan Hospital, Yatomi, Japan

2 Tokyo Institute of Technology, Tokyo, Japan

References

[1] Cruz-Jentoft AJ, Landi F, Schneider SM, et al. Prevalence of and interventions for sarcopenia in ageing adults: A systematic review. Report of the International Sarcopenia Initiative (EWGSOP and IWGS). Age and Ageing. 2014;**43**:748-759

[2] Iwashyna TJ. Survivorship will be the defining challenge of critical care in the 21st century. Annals of Internal Medicine. 2010;**153**:204-205

[3] Misak CJ. ICU-acquired weakness: Obstacles and interventions for rehabilitation. American Journal of Respiratory and Critical Care Medicine. 2011;**183**:845-846

[4] Leijten FS, Harinck-de Weerd JE, Poortvliet DC, et al. The role of polyneuropathy in motor convalescence after prolonged mechanical ventilation. The Journal of the American Medical Association. 1995;**274**:1221-1225

[5] Herridge MS, Cheung AM, Tansey CM, et al, for the Canadian Critical Care Trials Group. One-year outcomes in survivors of the acute respiratory distress syndrome. The New England Journal of Medicine. 2003;**348**:683-693

[6] Herridge MS, Tansey CM, Matte A, et al. Functional disability 5 years after acute respiratory distress syndrome. The New England Journal of Medicine. 2011;**364**:1293-1304

[7] Schefold JC, Bierbrauer J, Weber-Carstens S. Intensive care unit - acquired weakness (ICUAW) and muscle wasting in critically ill patients with severe sepsis and septic shock. Journal of Cachexia Sarcopenia and Muscle. 2010;**1**:147-157

[8] Schweickert WD, Kress JP. Implementing early mobilization interventions in mechanically ventilated patients in the ICU. Chest. 2011;**140**:1612-1617

[9] Stevens RD, Dowdy DW, Michaels RK, et al. Neuromuscular dysfunction acquired in critical illness: A systematic review. Intensive Care Medicine. 2007;**33**:1876-1891

[10] Latronico N, Bertolini G, Guarneri B, et al. Simplified electrophysiological evaluation of peripheral nerves in critically ill patients: The Italian multi-centre CRIMYNE study. Critical Care. 2007;**11**:R11

[11] Fenzi F, Latronico N, Refatti N, et al. Enhanced expression of E-selectin on the vascular endothelium of peripheral nerve in critically ill patients with neuromuscular disorders. Acta Neuropathologica. 2003;**106**:75-82

[12] Zink W, Kollmar R, Schwab S. Critical illness polyneuropathy and myopathy in the intensive care unit. Nature Reviews Neurology. 2009;**5**:372-379

[13] Latronico N, Fenzi F, Recupero D, et al. Critical illness myopathy and neuropathy. Lancet. 1996;**347**:1579-1582

[14] Bogdanski R, Blobner M, Werner C. Critical illness polyneuropathy and myopathy: Do they persist for lifetime? Critical Care Medicine. 2003;**31**:1279-1280

[15] Leijten FS, De Weerd AW, Poortvliet DC, et al. Critical illness polyneuropathy in multiple organ dysfunction syndrome and weaning from the ventilator. Intensive Care Medicine. 1996;**22**:856-861

[16] Hermans G, De Jonghe B, Bruyninckx F, et al. Clinical review: Critical illness polyneuropathy and myopathy. Critical Care. 2008;**12**:238

[17] Guarneri B, Bertolini G, Latronico N. Long-term outcome in patients with critical illness myopathy or neuropathy: The Italian multicenter CRIMYNE study. Journal of Neurology Neurosurgery and Psychiatry. 2008;**79**:838-841

[18] Bednarik J, Lukas Z, Vondracek P. Critical illness polyneuromyopathy: The electrophysiological components of a complex entity. Intensive Care Medicine. 2003;**29**:1505-1514

[19] Hund E. Neurological complications of sepsis: Critical illness polyneuropathy and myopathy. Journal of Neurology. 2001;**248**:929-934

[20] Tennilä A, Salmi T, Pettilä V, et al. Early signs of critical illness polyneuropathy in ICU patients with systemic inflammatory response syndrome or sepsis. Intensive Care Medicine. 2000;**26**:1360-1363

[21] Bercker S, Weber-Carstens, Deja M, et al. Critical illness polyneuropathy and myopathy in patients with acute respiratory distress syndrome. Critical Care Medicine 2005;**33**: 711-715

[22] de Letter MA, Schmitz PI, Visser LH, et al. Risk factors for the development of polyneuropathy and myopathy in critically ill patients. Critical Care Medicine. 2001;**29**:2281-2286

[23] De Jonghe B, Bastuji-Garin S, Sharshar T, et al. Does ICU-acquired paresis lengthen weaning from mechanical ventilation? Intensive Care Medicine. 2004;**30**:1117-1121

[24] Zifko UA. Long-term outcome of critical illness polyneuropathy. Muscle Nerve Suppl. 2000;**9**:S49-S52

[25] Fletcher SN, Kennedy DD, Ghosh IR, et al. Persistent neuromuscular and neurophysiologic abnormalities in long-term survivors of prolonged critical illness. Critical Care Medicine. 2003;**31**:1012-1016

[26] Intiso D, Amoruso L, Zarrelli M, et al. Long-term functional outcome and health status of patients with critical illness polyneuromyopathy. Acta Neurologica Scandinavica. 2011;**123**:211-219

[27] Latronico N, Shehu I, Seghelini E. Neuromuscular sequelae of critical illness. Current Opinion in Critical Care. 2005;**11**:381-390

[28] Garnacho-Montero J, Madrazo-Osuna J, Garcia-Garmendia JL, et al. Critical illness polyneuropathy: Risk factors and clinical consequences. A cohort study in septic patients. Intensive Care Medicine. 2001;**27**:1288-1296

[29] Latronico N, Peli E, Botteri M. Critical illness myopathy and neuropathy. Current Opinion in Critical Care. 2005;**11**:126-132

[30] de Jonghe B, Lacherade JC, Sharshar T, et al. Intensive care unit-acquired weakness: Risk factors and prevention. Critical Care Medicine. 2009;**37**:S309-S315

[31] Zochodne DW, Bolton CF, Wells GA, et al. Critical illness polyneuropathy. A complication of sepsis and multiple organ failure. Brain. 1987;**110**:819-841

[32] Carré JE, Orban JC, Re L, et al. Survival in critical illness is associated with early activation of mitochondrial biogenesis. American Journal of Respiratory and Critical Care Medicine. 2010;**182**:745-751

[33] Novak KR, Nardelli P, Cope TC, et al. Inactivation of sodium channels underlies reversible neuropathy during critical illness in rats. The Journal of Clinical Investigation. 2009;**119**:1150-1158

[34] Witt NJ, Zochodne DW, Bolton CF, et al. Peripheral nerve function in sepsis and multiple organ failure. Chest. 1991;**99**:176-184

[35] Friedrich O, Fink RH, Hund E. Understanding critical illness myopathy: approaching the pathomechanism. The Journal of Nutrition. 2005;**135**:1813S-1817S

[36] Dirks ML, Wall BT, Snijders T, et al. Neuromuscular electrical stimulation prevents muscle disuse atrophy during leg immobilization in humans. Acta Physiologica. 2014;**210**: 628-641

[37] Suetta C, Frandsen U, Jensen L, et al. Aging affects the transcriptional regulation of human skeletal muscle disuse atrophy. PLoS One. 2012;**7**:e51238

[38] Griffiths RD, Palmer TE, Helliwell T, et al. Effect of passive stretching on the wasting of muscle in the critically ill. Nutrition. 1995;**11**:428-432

[39] Jaber S, Petrof BJ, Jung B, et al. Rapidly progressive diaphragmatic weakness and injury during mechanical ventilation in humans. American Journal of Respiratory and Critical Care Medicine. 2011;**183**:364-371

[40] van den Berghe G, Schoonheydt K, Becx P, et al. Insulin therapy protects the central and peripheral nervous system of intensive care patients. Neurology. 2005;**64**:1348-1353

[41] De Jonghe B, Bastuji-Garin S, Durand MC, et al. Respiratory weakness is associated with limb weakness and delayed weaning in critical illness. Critical Care Medicine. 2007;**35**:2007-2015

[42] Demoule A, Jung B, Prodanovic H, et al. Diaphragm dysfunction on admission to the intensive care unit. Prevalence, risk factors, and prognostic impact—A prospective study. American Journal of Respiratory and Critical Care Medicine. 2013;**188**:213-219

[43] Zifko UA, Zipko HT, Bolton CF. Clinical and electrophysiological findings in critical illness polyneuropathy. Journal of Neurological Sciences. 1998;**159**:186-193

[44] Patel BK, Pohlman AS, Hall JB, et al. Impact of early mobilization on glycemic control and ICU-acquired weakness in critically ill patients who are mechanically ventilated. Chest. 2014;**146**:583-589

[45] Hermans G, De Jonghe B, Bruyninckx F, et al. Interventions for preventing critical illness polyneuropathy and critical illness myopathy. The Cochrane Database of Systematic Reviews. 2014;**1**:CD006832

[46] MacFarlane IA, Rosenthal FD. Severe myopathy after status asthmaticus. Lancet. 1977;**2**: 615

[47] Campellone JV, Lacomis D, Kramer DJ, et al. Acute myopathy after liver transplantation. Neurology 1998;**50**:46-53

[48] Massa R, Carpenter S, Holland P, et al. Loss and renewal of thick myofilaments in glucocorticoid-treated rat soleus after denervation and reinnervation. Muscle & Nerve. 1992;**15**:1290-1298

[49] Steinberg KP, Hudson LD, Goodman RB, et al. Efficacy and safety of corticosteroids for persistent acute respiratory distress syndrome. The New England Journal of Medicine. 2006;**354**:1671-1684

[50] Papazian L, Forel JM, Gacouin A, et al. Neuromuscular blockers in early acute respiratory distress syndrome. The New England Journal of Medicine. 2010;**363**:1107-1116

[51] Nanas S, Kritikos K, Angelopoulos E, et al. Predisposing factors for critical illness polyneuromyopathy in a multidisciplinary intensive care unit. Acta Neurologica Scandinavica. 2008;**118**:175-181

[52] Van den Berghe G, Wouters P, Weekers F, et al. Intensive insulin therapy in the critically ill patients. The New England Journal of Medicine. 2001;**345**:1359-1367

[53] Annane D, Bellissant E, Cavaillon J. Septic shock. Lancet. 2005;**365**:63-78

[54] Cohen J. The immunopathogenesis of sepsis. Nature. 2002;**420**:885-891

[55] Schefold JC, Hasper D, Volk HD, et al. Sepsis: Time has come to focus on the later stages. Medical Hypotheses. 2008;**71**:203-208

[56] Callahan LA, Supinski GS. Sepsis-induced myopathy. Critical Care Medicine. 2009;**37**: S354-S367

[57] Burnham EL, Moss M, Ziegler TR. Myopathies in critical illness: Characterization and nutritional aspects. The Journal of Nutrition. 2005;**135**:S1818-S1823

[58] Smith IJ, Lecker SH, Hasselgren PO. Calpain activity and muscle wasting in sepsis. American Journal of Physiology Endocrinology and Metabolism. 2008;**295**:E762-E771

[59] Bhattacharyya J, Thompson K, Sayeed MM. Calcium-dependent and calcium-independent protease activities is skeletal muscle during sepsis. Circulatory Shock. 1991;**35**:117-122

[60] Salazar JJ, Michele DE, Brooks SV. Inhibition of calpain prevents muscle weakness and disruption of sarcomere structure during hindlimb suspension. Journal of Applied Physiology. 2010;**108**:120-127

[61] Kraner SD, Wang Q, Novak KR, et al. Upregulation of the CaV 1.1-ryanodine receptor complex in a rat model of critical illness myopathy. American Journal of Physiology Regulatory Integrative and Comparative Physiology. 2011;**300**:R1384-R1391

[62] Wei W, Fareed MU, Evenson A, et al. Sepsis stimulates calpain activity in skeletal muscle by decreasing calpastatin activity but does not activate caspase-3. American Journal of Physiology Regulatory Integrative and Comparative Physiology. 2005;**288**:R580-R590

[63] Sandri M. Protein breakdown in muscle wasting: Role of autophagy-lysosome and ubiquitin-proteasome. The International Journal of Biochemistry & Cell Biology. 2013;**45**: 2121-2129

[64] Zhao J, Brault JJ, Schild A, et al. FoxO3 coordinately activates protein degradation by the autophagic/lysosomal and proteasomal pathways in atrophying muscle cells. Cell Metabolism. 2007;**6**:472-483

[65] Llano-Diez M, Gustafson AM, Olsson C, et al. Muscle wasting and the temporal gene expression pattern in a novel rat intensive care unit model. BMC Genomics. 2011;**12**:602

[66] Batt J, Bain J, Goncalves J, et al. Differential gene expression profiling of short and long term denervated muscle. FASEB Journal. 2006;**20**:115-117

[67] Ochala J, Renaud G, Llano Diez M, et al. Diaphragm muscle weakness in an experimental porcine intensive care unit model. PLoS One. 2011;**6**:e20558

[68] Constantin D, McCullough J, Mahajan RP, et al. Novel events in the molecular regulation of muscle mass in critically ill patients. Journal of Physiology. 2011;**589**:3883-3895

[69] Klaude M, Hammarqvist F, Wemerman J. An assay of microsomal membrane-associated proteasomes demonstrates increased proteolytic activity in skeletal muscle of intensive care unit patients. Clinical Nutrition. 2005;**24**:259-265

[70] Hooijman PE, Beishuizen A, Witt CC, et al. Diaphragm muscle fiber weakness and ubiquitin-proteasome activation in critically ill patients. American Journal of Respiratory and Critical Care Medicine. 2015;**191**:1126-1138

[71] Iida Y, Yamada S, Nishida O, et al. Body mass index is negatively correlated with respiratory muscle weakness and interleukin-6 production after coronary artery bypass grafting. Journal of Critical Care. 2010;**25**:172.e1-8

[72] Gorson KC. Approach to neuromuscular disorders in the intensive care unit. Neurocritical Care. 2005;**3**:195-212

[73] Hermans G, Vanhorebeek I, Derde S, et al. Metabolic aspects of critical illness polyneuromyopathy. Critical Care Medicine. 2009;**37**:S391-S397

[74] Ochala J, Ahlbeck K, Radell PJ, et al. Factors underlying the early limb muscle weakness in acute quadriplegic myopathy using an experimental ICU porcine model. PLoS One. 2011;**6**:e20876

[75] Kortebein P, Ferrando A, Lombeida J, et al. Effect of 10 days of bed rest on skeletal muscle in healthy older adults. The Journal of the American Medical Association. 2007;**297**:1772-1774

[76] Rich MM, Pinter MJ. Sodium channel inactivation in an animal model of acute quadriplegic myopathy. Annals of Neurology. 2001;**50**:26-33

[77] Rich MM, Pinter MJ. Crucial role of sodium channel fast inactivation in muscle fi breinexcitability in a rat model of critical illness myopathy. Journal of Physiology. 2003;**547**:555-566

[78] Filatov GN, Rich MM. Hyperpolarized shifts in the voltage dependence of fast inactivation of Nav1.4 and Nav1.5 in a rat model of critical illness myopathy. Journal of Physiology. 2004;**559**:813-820

[79] Khan J, Harrison TB, Rich MM. Mechanisms of neuromuscular dysfunction in critical illness. Critical Care Clinics. 2008;**24**:165-177

[80] Rossignol B, Gueret G, Pennec JP, et al. Effects of chronic sepsis on the voltage-gated sodium channel in isolated rat muscle fibers. Critical Care Medicine. 2007;**35**:351-357

[81] Klionsky DJ. Autophagy: From phenomenology to molecular understanding in less than a decade. Nature Reviews Molecular Cell Biology. 2007;**8**:931-937

[82] Mammucari C, Schiaffino S, Sandri M. Downstream of Akt: FoxO3 and mTOR in the regulation of autophagy in skeletal muscle. Autophagy. 2008;**4**:524-526

[83] Zhao J, Brault JJ, Schild A, et al. Coordinate activation of autophagy and the proteasome pathway by FoxO transcription factor. Autophagy. 2008;**4**:378-380

[84] Bertaggia E, Coletto L, Sandri M. Posttranslational modifications control FoxO3 activity during denervation. American Journal of Physiology Cell Physiology. 2012;**302**:C587-C596

[85] Masiero E, Agatea L, Mammucari C, et al. Autophagy is required to maintain muscle mass. Cell Metabolism. 2009;**10**:507-515

[86] Tiao G, Hobler S, Wang JJ, et al. Sepsis is associated with increased mRNAs of the ubiquitin - proteasome proteolytic pathway in human skeletal muscle. The Journal of Clinical Investigation. 1997;**99**:163-168

[87] Yousef AA, Suliman GA, Mabrouk MM. The value of correlation of serum 20S proteasome concentration and percentage of lymphocytic apoptosis in critically ill patients: A prospective observational study. Critical Care. 2010;**14**:R215

[88] Jespersen JG, Nedergaard A, Reitelseder S, et al. Activated protein synthesis and suppressed protein breakdown signaling in skeletal muscle of critically ill patients. PLoS One. 2011;**6**:e18090

[89] de Boer MD, Selby A, Atherton P, et al. The temporal responses of protein synthesis, gene expression and cell signalling in human quadriceps muscle and patellar tendon to disuse. Journal of Physiology. 2007;**585**:241-251

[90] Léger B, Senese R, Al-Khodairy AW, et al. Atrogin-1, MuRF1, and FoXO, as well as phosphorylated GSK-3beta and 4E-BP1 are reduced in skeletal muscle of chronic spinal cord-injured patients. Muscle & Nerve. 2009;**40**:69-78

[91] Hussain SN, Mofarrahi M, Sigala I, et al. Mechanical ventilation – induced diaphragm disuse in humans triggers autophagy. American Journal of Respiratory and Critical Care Medicine. 2010;**182**:1377-1386

[92] Vanhorebeek I, Gunst J, Derde S, et al. Insufficient activation of autophagy allows cellular damage to accumulate in critically ill patients. The Journal of Clinical Endocrinology and Metabolism. 2011;**96**:E633-E645

[93] Sakuma K, Aoi W, Yamaguchi A. The intriguing regulators of muscle mass in sarcopenia and muscular dystrophy. Frontiers in Aging Neuroscience. 2014;**6**:230

[94] McClung JM, Judge AR, Powers SK, et al. p38 MARK links oxidative stress to autophagy-related gene expression in cachectic muscle wasting. American Journal of Physiology Cell Physiology. 2010;**298**:C542-549

[95] O'Leary MF, Hood DA. Effect of prior chronic contractile activity on mitochondrial function and apoptotic protein expression in denervated muscle. Journal of Applied Physiology. 2008;**105**:114-120

[96] Sakuma K, Kinoshita M, Ito Y, et al. p62/SQSTM1 but not LC3 is accumulated in sarcopenic muscle of mice. Journal of Cachexia, Sarcopenia and Muscle. 2016;**7**:204-212

[97] Rennie MJ. Anabolic resistance in critically ill patients. Critical Care Medicine. 2009;**37**: S398-S399

[98] Thomason DB, Biggs RB, Booth FW. Protein metabolism and beta-myosin heavy-chain mRNA in unweighted soleus muscle. The American Journal of Physiology. 1989;**257**:R300-R305

[99] Stitt TN, Drujan D, Clarke BA, et al. The IGF-1/PI3K/Akt pathway prevents expression of muscle atrophy-induced ubiquitin ligases by inhibiting FOXO transcription factors. Molecular Cell. 2004;**14**:395-403

[100] Krawiec BJ, Frost RA, Vary TC, et al. Hindlimb casting decreases muscle mass in part by proteasome-dependent proteolysis but independent of protein synthesis. American Journal of Physiology Endocrinology and Metabolism. 2005;**289**:E969-E980

[101] Lang CH, Frost RA, Vary TC. Regulation of muscle protein synthesis during sepsis and inflammation. American Journal of Physiology Endocrinology and Metabolism. 2007;**293**:E453-E459

[102] Finfer S, Chittock DR, Su SY, et al. Intensive versus conventional glucose control in critically ill patients. The New England Journal of Medicine. 2009;**360**:1283-1297

[103] Van den Berghe G, Wilmer A, Hermans G, et al. Intensive insulin therapy in the medical ICU. The New England Journal of Medicine. 2006;**354**:449-461

[104] Vlasselaers D, Milants I, Desmet L, et al. Intensive insulin therapy for patients in paediatric intensive care: a prospective, randomised controlled study. Lancet. 2009;**373**:547-556

[105] Dellinger RP, Levy MM, Rhodes A, et al. Surviving Sepsis Campaign: International guidelines for management of severe sepsis and septic shock, 2012. Intensive Care Medicine. 2013;**39**:165-228

[106] Weber-Carstens S, Schneider J, Wollersheim T, et al. Critical illness myopathy and GLUT4: Significance of insulin and muscle contraction. American Journal of Respiratory and Critical Care Medicine. 2013;**187**:387-396

[107] Mesotten D, Swinnen JV, Vanderhoydonc F, et al. Contribution of circulating lipids to the improved outcome of critical illness by glycemic control with intensive insulin therapy. The Journal of Clinical Endocrinology and Metabolism. 2004;**89**:219-226

[108] Hermans G, Wilmer A, Meersseman W, et al. Impact of intensive insulin therapy on neuromuscular complications and ventilator-dependency in MICU. American Journal of Respiratory and Critical Care Medicine. 2007;**175**:480-489

[109] Hermans G, Schrooten M, Van Damme P, et al. Benefits of intensive insulin therapy on neuromuscular complications in routine daily critical care practice: A retrospective study. Critical Care. 2009;**13**:R5

[110] Derde S, Hermans G, Derese I, et al. Muscle atrophy and preferential loss of myosin in prolonged critically ill patients. Critical Care Medicine. 2012;**40**:79-89

[111] Van den Berghe G, Schetz M, Vlasselaers D, et al. Clinical review: Intensive insulin therapy in critically ill patients: NICESUGAR or Leuven blood glucose target? The Journal of Clinical Endocrinology and Metabolism. 2009;**94**:3163-3170

[112] Chambers MA, Moylan JS, Reid MB. Physical inactivity and muscle weakness in the critically ill. Critical Care Medicine. 2009;**37**:S337-S346

[113] Barr J, Fraser GL, Puntillo K, et al. Clinical practice guidelines for the management of pain, agitation, and delirium in adult patients in the intensive care unit. Critical Care Medicine. 2013;**41**:263-306

[114] Schweickert WD, Pohlman MC, Pohlman AS, et al. Early physical and occupational therapy in mechanically ventilated, critically ill patients: A randomised controlled trial. Lancet. 2009;**373**:1874-1882

[115] Lipshutz AK, Gropper MA. Acquired neuromuscular weakness and early mobilization in the intensive care unit. Anesthesiology. 2013;**118**:202-215

[116] Pohlman MC, Schweickert WD, Pohlman AS, et al. Feasibility of physical and occupational therapy beginning from initiation of mechanical ventilation. Critical Care Medicine. 2010;**38**:2089-2094

[117] Burtin C, Clerckx B, Robbeets C, et al. Early exercise in critically ill patients enhances short-term functional recovery. Critical Care Medicine. 2009;**37**:2499-2505

[118] Lee CM, Fan E. ICU-acquired weakness: What is preventing its rehabilitation in critically ill patients? BMC Medicine. 2012;**10**:115

[119] Gosselink R, Bott J, Johnson M, et al. Physiotherapy for adult patients with critical illness: Recommendations of the European Respiratory Society and European Society of Intensive Care Medicine Task Force on Physiotherapy for Critically Ill Patients. Intensive Care Medicine. 2008;**34**:1188-1199

[120] Winkelman C, Higgins PA, Chen YJ. Activity in the chronically critically ill. Dimensions of Critical Care Nursing. 2005;**24**:281-290

[121] Zanni JM, Korupolu R, Fan E, et al. Rehabilitation therapy and outcomes in acute respiratory failure: An observational pilot project. Journal of Critical Care. 2010;**25**:254-262

[122] Hopkins RO, Spuhler VJ, Thomsen GE. Transforming ICU culture to facilitate early mobility. Critical Care Clinics. 2007;**23**:81-96

[123] Martin UJ, Hincapie L, Nimchuk M, et al. Impact of whole-body rehabilitation in patients receiving chronic mechanical ventilation. Critical Care Medicine. 2005;**33**:2259-2265

[124] Connolly B, Denehy L, Brett S, et al. Exercise rehabilitation following hospital discharge in survivors of critical illness: An integrative review. Critical Care. 2012;**16**:226

[125] Gibson JN, Smith K, Rennie MJ. Prevention of disuse muscle atrophy by means of electrical stimulation: Maintenance of protein synthesis. Lancet. 1988;**2**:767-770

[126] Rodriguez PO, Setten M, Maskin LP, et al. Muscle weakness in septic patients requiring mechanical ventilation: Protective effect of transcutaneous neuromuscular electrical stimulation. Journal of Critical Care. 2012;**27**:319.e1- e8

[127] Poulsen JB, Moller K, Jensen CV, et al. Effect of transcutaneous electrical muscle stimulation on muscle volume in patients with septic shock. Critical Care Medicine. 2011;**39**:456-461

[128] Routsi C, Gerovasili V, Vasileiadis I, et al. Electrical muscle stimulation prevents critical illness polyneuromyopathy: A randomized parallel intervention trial. Critical Care. 2010;**14**:R74

[129] Zanotti E, Felicetti G, Maini M, et al. Peripheral muscle strength training in bedbound patients with COPD receiving mechanical ventilation: Effect of electrical stimulation. Chest 2003;**124**:292-296

[130] Abdellaoui A, Prefaut C, Gouzi F, et al. Skeletal muscle effects of electrostimulation after COPD exacerbation: A pilot study. The European Respiratory Journal. 2011;**38**:781-788

[131] Sillen MJ, Speksnijder CM, Eterman RM, et al. Effects of neuromuscular electrical stimulation of muscles of ambulation in patients with chronic heart failure or COPD: A systematic review of the English-language literature. Chest. 2009;**136**:44-61

[132] Gruther W, Kainberger F, Fialka-Moser V, et al. Effects of neuromuscular electrical stimulation on muscle layer thickness of knee extensor muscles in intensive care unit patients: A pilot study. Journal of Rehabilitation Medicine. 2010;**42**:593-597

http://dx.doi.org/10.5772/intechopen.69051

[133] Gerovasili V, Stefanidis K, Vitzilaios K, et al. Electrical muscle stimulation preserves the muscle mass of critically ill patients: A randomized study. Critical Care. 2009;**13**:R161

[134] Bloch S, Polkey MI, Griffiths M, et al. Molecular mechanisms of intensive care unit-acquired weakness. The European Respiratory Journal. 2012;**39**:1000-1011

[135] Gerovasili V, Tripodaki E, Karatzanos E, et al. Short-term systemic effect of electrical muscle stimulation in critically ill patients. Chest. 2009;**136**:1249-1256

[136] Angelopoulos E, Karatzanos E, Dimopoulos S, et al. Acute microcirculatory effects of medium frequency versus high frequency neuromuscular electrical stimulation in critically ill patients—A pilot study. Annals of Intensive Care. 2013;**3**:39

[137] Hamada T, Sasaki H, Hayashi T, et al. Enhancement of whole body glucose uptake during and after human skeletal muscle low-frequency electrical stimulation. Journal of Applied Physiology. 2003;**94**:2107-2112

[138] Gordon JW, Rungi AA, Inagaki H, et al. Effects of contractile activity on mitochondrial transcription factor A expression in skeletal muscle. Journal of Applied Physiology. 2001;**90**:389-396

[139] Bouletreau P, Patricot MC, Saudin F, et al. Effects of intermittent electrical stimulations on muscle catabolism in intensive care patients. Journal of Parenteral and Enteral Nutrition. 1987;**11**:552-555

[140] Iwatsu K, Iida Y, Kono Y, et al. Neuromuscular electrical stimulation may attenuate muscle proteolysis after cardiovascular surgery: A preliminary study. The Journal of Thoracic and Cardiovascular Surgery. 2017;**153**:373-379

[141] Karatzanos E, Gerovasili V, Zervakis D, et al. Electrical muscle stimulation: An effective form of exercise and early mobilization to preserve muscle strength in critically ill patients. Critical Care Research and Practice. 2012;**2012**:432752

[142] Watters JM, Kirkpatrick SM, Norris SB, et al. Immediate postoperative enteral feeding results in impaired respiratory mechanics and decreased mobility. Annals of Surgery. 1997;**226**:369-377

[143] Casaer MP, Mesotten D, Hermans G, et al. Early versus late parenteral nutrition in critically ill adults. The New England Journal of Medicine. 2011;**365**:506-517

[144] Hermans G, Casaer MP, Clerckx B, et al. Effect of tolerating macronutrient deficit on the development of intensive-care unit acquired weakness: A subanalysis of the EPaNic trial. The Lancet Respiratory Medicine. 2013;**1**:621-629

[145] Doig GS, Simpson F, Sweetman EA, et al. Early parenteral nutrition in critically ill patients with short-term relative contraindications to early enteral nutrition: A randomized controlled trial. The Journal of the American Medical Association. 2013;**309**:2130-2138

[146] National Heart, Lung, and Blood Institute Acute Respiratory Distress Syndrome Clinical Trials, Rice TW, Wheeler AP, Thompson BT, et al. Initial trophic vs full enteral feeding in patients with acute lung injury: The EDEN randomized trial. The Journal of the American Medical Association. 2012;**307**:795-803

[147] Needham DM, Dinglas VD, Bienvenu OJ, et al. One year outcomes in patients with acute lung injury randomised to initial trophic or full enteral feeding: prospective follow-up of EDEN randomised trial. British Medical Journal. 2013;**346**:f1532

[148] Wernerman J. Glutamine supplementation. Annals Intensive Care. 2011;**1**:25

[149] Novak F, Heyland DK, Avenell A, et al. Glutamine supplementation in serious illness: A systematic review of the evidence. Critical Care Medicine. 2002;**30**:2022-2029

[150] Andrews PJ, Avenell A, Noble DW, et al. Randomised trial of glutamine, selenium, or both, to supplement parenteral nutrition for critically ill patients. British Medical Journal. 2011;**342**:d1542

[151] Heyland D, Muscedere J, Wischmeyer PE, et al. A randomized trial of glutamine and antioxidants in critically ill patients. The New England Journal of Medicine. 2013;**368**:1489-1497

[152] Manzanares W, Dhaliwal R, Jiang X, et al. Antioxidant micronutrients in the critically ill: A systematic review and meta-analysis. Critical Care. 2012;**16**:R66

[153] Visser J, Labadarios D, Blaauw R. Micronutrient supplementation for critically ill adults: A systematic review and meta-analysis. Nutrition. 2011;**27**:745-758

[154] Powers SK, Hudson MB, Nelson WB, et al. Mitochondria-targeted antioxidants protect against mechanical ventilation-induced diaphragm weakness. Critical Care Medicine. 2011;**39**:1749-1759

[155] Casaer MP, Van den Berghe G. Nutrition in the acute phase of critical illness. The New England Journal of Medicine. 2014;**370**:1227-1236

[156] Sakuma K, Yamaguchi A. Molecular mechanisms in aging and current strategies to counteract sarcopenia. Current Aging Science. 2010;**3**:90-101